Astronomers' Observing Guides

For further volumes:
http://www.springer.com/series/5338

Richard Schmude, Jr.

Artificial Satellites and How to Observe Them

Springer

Richard Schmude, Jr.
109 Tyus Street
Barnesville, GA, USA

ISSN 1611-7360
ISBN 978-1-4614-3914-1 ISBN 978-1-4614-3915-8 (eBook)
DOI 10.1007/978-1-4614-3915-8
Springer NewYork Heidelberg Dordrecht London

Library of Congress Control Number: 2012939432

Printed on acid-free paper

Springer is part of Springer Science+Business Media (www.springer.com)

This book is dedicated to the many people who have helped me along the way: first, to my father and mother, Richard and Winifred Schmude, who showed me the stars and answered all my questions; to the many fine teachers, professors, and school administrators who guided me during my childhood and early adulthood; to the members of the Astronomical League, the Association of Lunar and Planetary Observers, and the Royal Astronomical Society of Canada who supported and encouraged me; to the librarians at Gordon College who helped me obtain critical information for this book; and, finally, to the webmasters who prepared and maintain the websites listed in the book.

Preface

I became interested in astronomy initially when I saw what appeared to be a countless number of stars from my parent's home in Cabin John, Maryland. I was no older than six when I had this life-changing view of the night sky. I purchased my first telescope at age 15 and shared it with my siblings and a neighbor girl, Kathy. This was my first experience with public outreach.

One of my early views of an artificial satellite was of the International Space Station in about the year 2000. I remember watching it grow brighter and reaching a peak brightness directly overhead.

This book is divided into two parts. The first part (Chaps. 1, 2, 3) summarizes our current knowledge of artificial satellites. I have chosen to focus on satellites designed to collect scientific data of the Earth (Chap. 3) and of other bodies (Chap. 2). In the second part (Chaps. 4, 5, 6), I describe how to observe, image, and study satellites. The last chapter describes how to compute the eclipse times of geostationary satellites.

Various websites are listed in this book, some of which may be updated or discontinued by the time this book is published. Changes in website addresses are bound to happen. What I have seen, though, is that a discontinued website is often replaced by something even better.

About the Author

Dr. Richard Schmude, Jr., was born in Washington, D. C., and attended public schools in Cabin John, Maryland; Los Angeles, California; and Houston, Texas. He started his college career at North Harris County College and graduated from Texas A&M University with a Bachelor of Arts degree in Chemistry. Later, he obtained a Master of Science degree in Chemistry, a Bachelor of Arts degree in Physics, and a Ph.D. in Physical Chemistry, all from Texas A&M University. He worked at NALCO Chemical Company as a graduate co-op student and at Los Alamos National Laboratory as a graduate research assistant.

Since 1994, Richard has taught astronomy, chemistry, and other science courses at Gordon College, Barnesville, Georgia. He is a tenured professor at this college and continues to teach his students (and others) in these areas. He has published over 100 scientific papers in many different journals and has given over 500 talks and conducted telescope viewing sessions and workshops for over 25,000 people.

Contents

Chapter 1

Satellite Basics

The former Union of Soviet Socialist Republics (USSR) launched the first artificial satellite on October 4, 1957. In order to launch this object, Soviet engineers had to overcome several obstacles including gravity, the proper functioning of the rocket, air resistance, tracking and insertion of the satellite into orbit. Since 1957, several nations have launched satellites. These instruments are used for a wide variety of applications that include space studies, communication, military, agriculture, weather observation and land/ocean studies. The first three chapters describe the different spacecraft observing both Earth and bodies beyond Earth. Chapters 4, 5, and 6 describe how one may observe, image and study satellites.

In this chapter, we focus on the launch and operation of a satellite. Accordingly, it is broken into eight parts, which are the atmosphere, rockets, launch and operation of a satellite, electromagnetic radiation and the Doppler effect, satellite communication and tracking, satellite lifetime, satellite visibility and satellite applications.

The Atmosphere

Earth is surrounded by a blanket of gas called the atmosphere. Gravity keeps it in place. Without gravity, it would spread out into space. Furthermore, if Earth's gravity were one sixth of what it is as on our Moon, most of the air we breathe probably would have escaped long ago. Earth's higher gravity is why we have an atmosphere whereas our Moon lacks one.

Gravity is why our atmosphere exerts pressure. At sea level, gases above exert a pressure of 1.0 atm. This is equivalent to 14.7 lb per square inch, or 760 Torr. The pressure fluctuates by a few tenths of a percent depending on weather conditions. In rare cases, such as in a tornado or a hurricane, the pressure may drop several percent.

Our atmosphere changes with altitude. Both the pressure and density drop with increasing altitude. Table 1.1 lists both the pressure and density in different parts of the troposphere.

Our atmosphere has four layers. The lowest one, the troposphere, extends from Earth's surface to an altitude of between 7 and 18 km. The altitude depends on the latitude. The troposphere extends to 18 km at the equator but only 7 km at a pole. The temperature almost always falls with increasing altitude in this layer.

R. Schmude, Jr., *Artificial Satellites and How to Observe Them*, Astronomers' Observing Guides, DOI 10.1007/978-1-4614-3915-8_1, © Springer Science+Business Media New York 2012

Table 1.1. Atmospheric pressure and density at different altitudes near the equator. The values were computed using a surface temperature of 20°C, a lapse rate of 7.0°C per km Whitaker (1996): 24 and a scale height of 7.0 km Goody and Walker (1972): 8

Altitude (km)	Pressure (atm)	Density (kg/m^3)
0	1.00	1.20
1	0.89	1.07
2	0.79	0.95
4	0.62	0.76
6	0.49	0.60
8	0.39	0.48
10	0.30	0.39
15	0.17	0.19

Fig. 1.1. This is an image of the troposphere and stratosphere taken from a balloon at an altitude of 25 km. The Hungarian Vega Astronomical Association (VAA) launched this balloon on October 31, 2011 (Courtesy of the VAA).

The troposphere is where storms, hurricanes, tornados and most clouds occur. Above it is the stratosphere, where temperatures rise with increasing altitude. It extends to about 50 km above sea level. Figure 1.1 was taken from a balloon at an altitude of 25 km. It shows the bluish troposphere and the black stratosphere. The pressure in this layer ranges from 0.3 to 0.001 atm.

Above the stratosphere lies the mesosphere. Temperatures fall with increasing altitude here. The air is about 10,000 times thinner in the mesosphere than it is at sea level. In spite of this, the air is thick enough to cause meteors to burn up. In fact the *Columbia* space shuttle broke apart in 2003 when it was passing through the mesosphere. Noctilucent clouds occur at an altitude of around 80 km, which is near the upper boundary of this layer.

The thermosphere is the highest layer. Temperatures rise with increasing altitude here. The gas density essentially reaches zero at an altitude of 800 km.

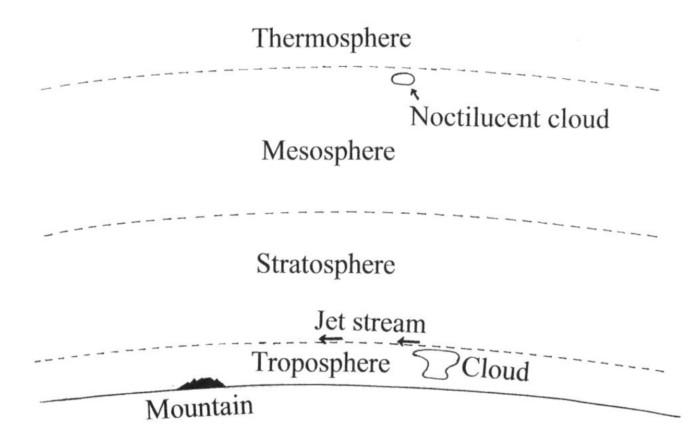

Fig. 1.2. The four layers of Earth's atmosphere at a latitude of 45°N. The approximate altitudes of the *troposphere, stratosphere, mesosphere* and *thermosphere* at 45°N are 0–13 km, 13–50 km, 50–85 km and above 85 km, respectively (Credit: Richard Schmude, Jr.).

Many satellites orbit Earth and pass through the tenuous thermosphere. Figure 1.2 illustrates the layers of the atmosphere. The temperature of the thermosphere may be 200°C hotter in the daytime than at night. This change affects the pressure and density. Temperature changes may also affect the orbits and lifetimes of satellites. This is discussed later.

The ionosphere contains lots of charged particles. It lies 100–300 km above sea level. Ham radio operators utilize this region when they communicate with those on other continents. Essentially, radio waves bounce off the ionosphere.

Air Resistance

Why does a feather fall slower than a rock? The answer is air resistance. Although the feather and rock feel gravitational pull, the rock falls faster because air resistance does not affect the rock's movement as much as the feather's movement. Air resistance is a frictional force between a moving object and air. It depends on several factors, including the object's speed and surface area. One may write the following for a falling object:

$$\text{Downward net force} = \text{gravitational force} - \text{air resistance force} \qquad (1.1)$$

In many cases, the gravitational force will equal the air resistance force. When this happens, the downward net force will equal zero, and the object will fall at a constant velocity.

Table 1.2 lists the measured acceleration of several objects dropped from a height of 2 m. Recall that acceleration is the change in velocity divided by the time interval. Without air resistance, the measured acceleration for all objects would be 9.81 m/s^2. As it turns out, all of the objects fall with accelerations less than this amount. Air resistance causes this. Figure 1.3 shows the distance, in meters, that objects fall in 0.64 s. The hypothetical object K has no air resistance and it falls farther than the others. The object experiencing the most air resistance (the balloon) falls the least amount.

Table 1.2. Measured acceleration of several objects when dropped from a height of 2 m. Chad Davies and the author collected all data

Object	Acceleration (m/s^2)	Diameter (cm)	Weight (N)
Baseball	9.67	7.1	1.4
Softball	9.42	9.1	1.7
Basketball	9.26	23	4.0
Vinyl ball	8.82	13	0.5
Styrofoam ball	8.49	14	0.5
Balloon	3.41[a]	16	0.031

[a]For the first 0.4 s. Afterwards the balloon traveled at its terminal velocity, which was 1.87 m/s

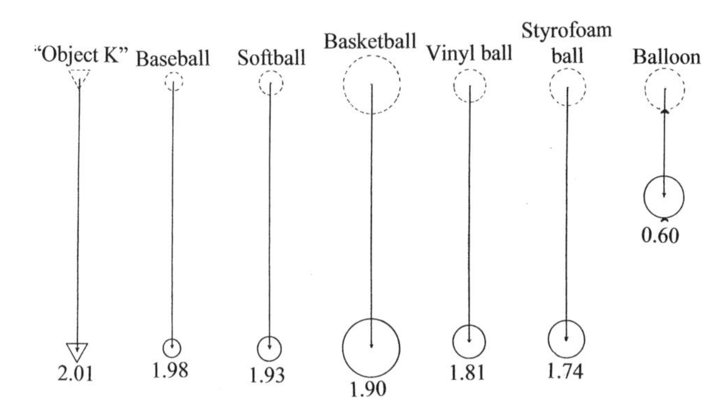

Fig. 1.3. The numbers at the bottom are the distance, in meters, which several objects fall after 0.64 s. Object K falls at an acceleration rate of 9.81 m/s^2 and does not experience any air resistance. These results are based on experimental data that Chad Davies and the author collected.

As a rocket rises, its velocity increases, its mass decreases and the air density decreases. This will cause the air resistance to change. In a typical launch, the rocket velocity increases and, hence, the air resistance force increases. Once the rocket is high enough, the air thins out enough for the air resistance force to drop in spite of the higher velocity. The air at an altitude of 200 km is so thin that air resistance is negligible.

Rockets

Have you ever watched a rocket launch? If so, do you think that it was moving at an initial speed of 28,000 km/h? (This is the speed needed to launch a satellite into orbit.) The answer is no. When a rocket lifts off its speed is much slower than this. It increases with time. Eventually, it reaches the speed needed to place a satellite in orbit.

The Saturn V rocket, used for the *Apollo 11* Moon mission, started off slowly. After 8 s, it was only 120 m above the ground. During this time it was moving at an average speed of 55 km/h. It also had an average acceleration of 4 m/s^2. The air resistance force was small during the first few seconds after liftoff. The speed

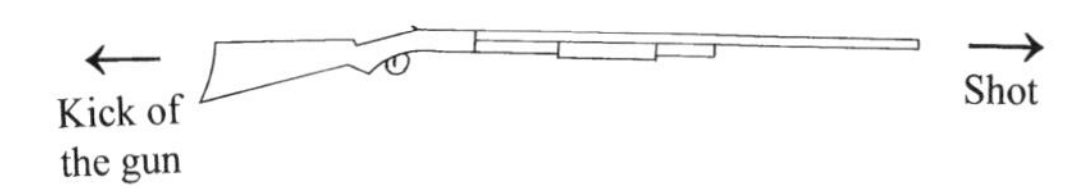

Fig. 1.4. When a shotgun is fired, the shot is propelled out of the barrel, and at the same time, the gun pushes back, which is its kick (Credit: Richard Schmude, Jr.).

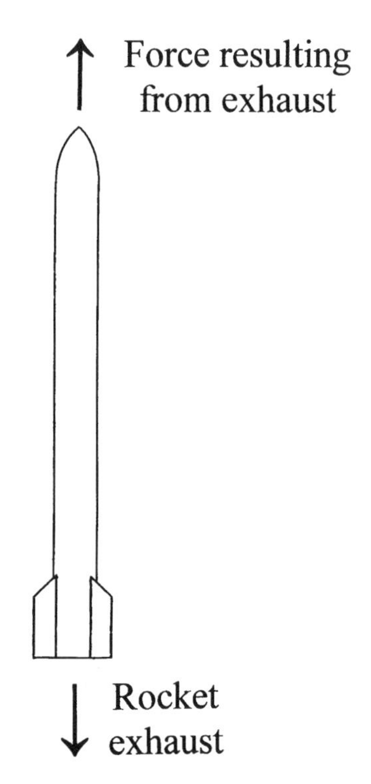

Fig. 1.5. When exhaust is pushed in one direction, the rocket moves in the opposite direction (Credit: Richard Schmude, Jr.).

of the rocket continued to increase, however, as it rose into space. Once it was high enough, its speed increased to the point where it was able to launch the spacecraft at the proper velocity. In essence, air resistance retarded the rocket's upward movement. To make matters more complicated, the air resistance depended on the air temperature, pressure and composition. It is for this reason that during the launch of the Saturn V, engineers monitored its velocity and acceleration. If adjustments needed to be made, the appropriate commands would have been sent to the crew.

What causes a rocket to move? One common misconception is that the exhaust pushes against the ground. This is not what happens. Instead it moves upward because of Newton's Third Law of Motion, which states: "For every action, there is an equal and opposite reaction." This is why a gun kicks after being fired. See Fig. 1.4. If one wants to launch a rocket straight up it is necessary to orient it so that it exerts a force straight down. Essentially, when material exits in one direction, the rocket moves in the other direction. See Fig. 1.5.

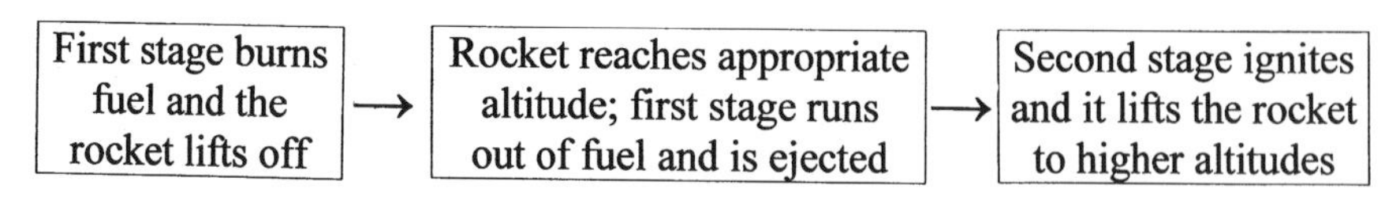

Fig. 1.6. The sequence of events for launching a two-stage rocket (Credit: Richard Schmude, Jr.).

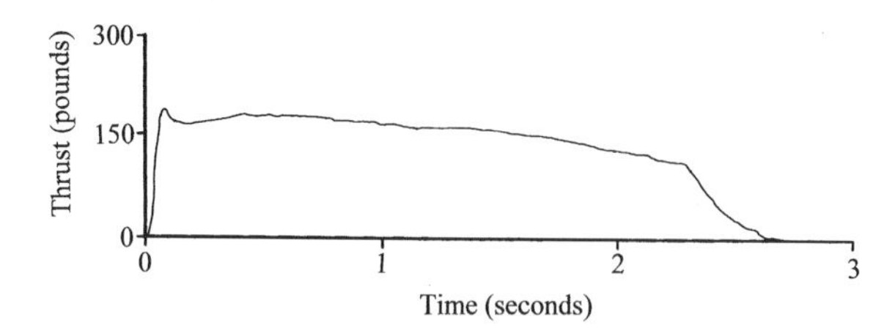

Fig. 1.7. A typical thrust curve for the K550W model rocket engine (Courtesy of Aerotech Division of RCS Rocket Motor Components Inc.).

The thrust is the force a rocket engine exerts. With a thrust of 5,000 N it may lift 5,000 N. Keep in mind though that most of this 5,000 N includes the weight of the launch vehicle and fuel. Therefore, if a rocket and its fuel weigh 4,000 N and the thrust is 5,000 N the maximum weight that may be lifted is 1,000 N.

Many large rockets have two, three or four stages. The advantage of this is that dead weight can be expelled. Figure 1.6 illustrates the basic launch process for a two-stage rocket.

Have you ever tried to balance an object on top of your head? Essentially, you would hold your head steady, and if the object began to move to the left you would move to prevent the object from moving further. A rocket works in the same way. In an ideal case, it will move straight up initially and afterwards will arc. Once it is high enough, its movement will be nearly parallel to Earth's surface. In most cases, it shifts from the planned path. Winds, air turbulence and changes in the mass distribution are responsible for this. Because of small rocket movements, operators should steer it. This is accomplished with gimbals and thrusters. A few rockets also have fins near their base. These also serve as a source of stability.

Model Rocket Launch

Let's think about the launch of a 9.5 lb fiberglass model rocket. It contains an altimeter and other equipment that are able to record the velocity, acceleration, altitude and barometric pressure during flight. It has a solid propellant engine (K550W rocket engine). Figure 1.7 shows a typical thrust curve. One pound of thrust equals 4.45 N. The rocket is loaded with its engine and placed on the launch site. It is aligned with a steel rod. After it is in place, a small igniter is attached. Once everything is in place, the field is cleared and the countdown starts: five, four, three, two, one, the igniter is lit and after about a second the rocket shoots up. See Fig. 1.8. Note the hot gas coming from the bottom. Since the gas moves downward, the rocket moves upward. During the first 2 m of flight, it moves up the rod. Figure 1.9 shows the altitude at different times. The engine burns for about 3 s. During the first 2.3 s, it exerts a thrust of between 120 and 180 lb or 530 and 800 N. During the next 0.7 s,

Fig. 1.8. Liftoff of a model rocket. During the first 2 m of its journey it moved up the metal post. This helped it move upwards (Credit: Chris Keir).

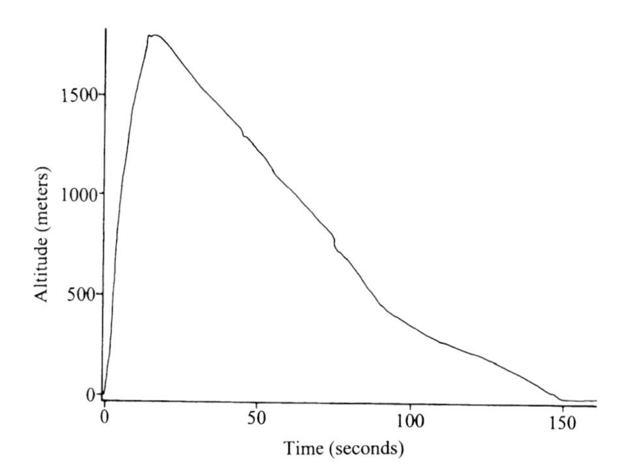

Fig. 1.9. The altitude versus time plot for the rocket shown in Fig. 1.8 with the K550W model rocket engine (Courtesy of Chris Keir and Richard Schmude, Jr.).

the thrust drops. After 3.0 s, the fuel runs out. At this point, the engine shuts down and the rocket is moving nearly straight up at a speed of 960 km/h. It continues to move upwards for another 10 s or so. Air resistance causes the upward velocity to drop. Once it reaches an altitude of about 1,800 m its upward velocity is zero, and it begins falling. Since this rocket does not attain a speed of 28,000 km/h it is unable to orbit Earth. Shortly after it begins falling, a parachute opens. Afterwards it falls at a nearly constant speed of 66 km/h.

Figure 1.10 shows the atmospheric pressure near the rocket. The pressure equaled 0.77 atm at the maximum altitude. This was much lower than on the ground (0.97 atm).

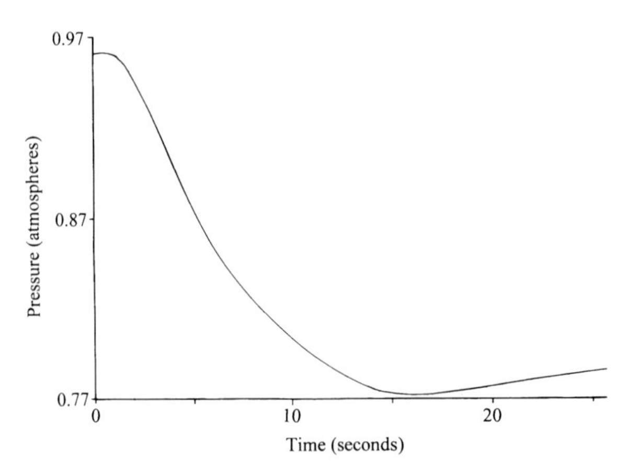

Fig. 1.10. The atmospheric pressure the rocket experienced during the first 26 s of flight. It reached a maximum altitude about 15 s after liftoff. This is also the time when the atmospheric pressure was lowest (Courtesy of Chris Keir and Richard Schmude, Jr.).

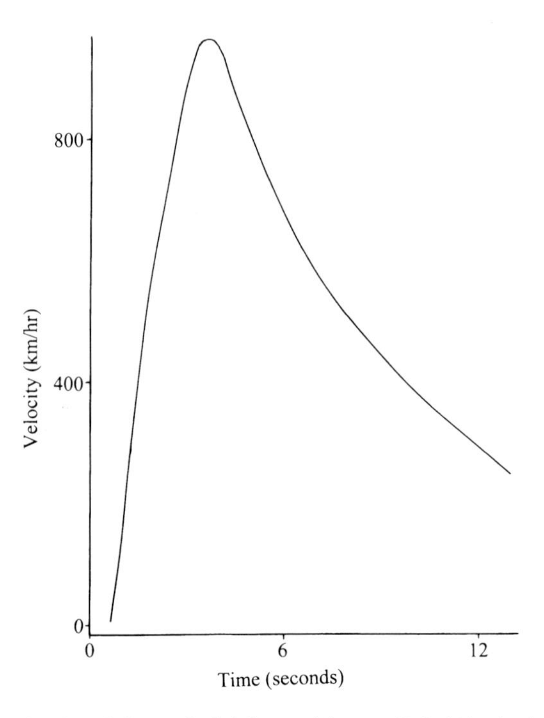

Fig. 1.11. The rocket velocity during the first 12 s of its flight (Courtesy of Chris Keir and Richard Schmude, Jr.).

Figure 1.11 shows the rocket velocity versus time. Engine thrust causes the velocity to increase during the first 3 s of flight. Afterwards, air resistance causes the velocity to fall. Table 1.3 lists the velocities, accelerations and altitudes at different times in the flight.

Larger rockets work in the same way as smaller ones. Both require thrust in order to accelerate. If they do not reach a speed of over 28,000 km/h, they fall back. Air resistance and wind also affect the trajectory regardless of rocket size.

Table 1.3. Velocity, acceleration and altitude of a model rocket launched on October 10, 2010. Chris Keir collected all data

Elapsed time (s)	Velocity (km/h)	Acceleration (m/s^2)	Altitude (m)
0	0	0	0
0.5	240	150	46
1.0	470	130	110
1.5	640	110	200
2.0	800	94	300
2.5	920	67	430
3.0	960	9.8	550
4.0	840	~0	760
6.0	610	~0	1,100

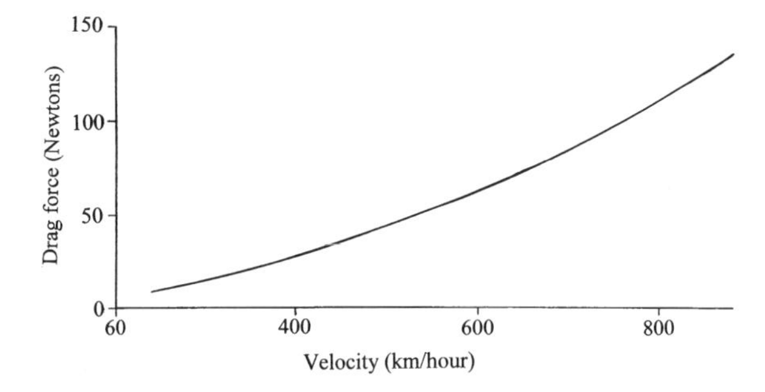

Fig. 1.12. The drag force caused by air resistance is plotted against the rocket velocity. The same rocket used in Fig. 1.8 is shown here (Courtesy of Chris Keir and Richard Schmude, Jr.).

The model rocket failed to reach higher altitudes because it ran out of fuel. If the engine could have burned for 1 min with 667 N of thrust, the rocket would have reached a velocity of 28,000 km/h in the absence of air resistance.

Figure 1.12 shows how the force of air resistance (or the drag force) changes for different velocities. At a velocity of 750 km/h, the drag force reaches 100 N, or 22.5 lb.

How does air resistance affect the rocket trajectory? A quick look back at Fig. 1.7 shows that the average thrust of the engine was about 150 lb (667 N). The rocket and its fuel had a mass of 4.5 kg at liftoff, and it may have had a mass of 4.1 kg after 2 s in flight. Therefore, its average mass was near 4.3 kg during its first 2 s of flight. Using an equation from physics (force = mass × acceleration, or acceleration = force ÷ mass) one may compute the acceleration (assuming no air resistance) as

$$\text{rocket acceleration} = 667\ \text{kg m / s}^2 \div 4.3\ \text{kg} = 155\ \text{m / s}^2 \qquad (1.2)$$

One may also compute the velocity after 2 s from Eq. 1.2 (assuming no air resistance) as

$$\text{velocity} = \text{initial velocity} + \text{acceleration} \times \text{time} \qquad (1.3)$$

In this equation the initial velocity equals zero, the acceleration equals 155 m/s^2 and the time is the elapsed time since liftoff, which in our case is 2.0 s. The velocity is computed as:

$$\text{velocity} = 0 + 155\,\text{m/s}^2 \times 2.0\,\text{s} = 310\,\text{m/s} \tag{1.4}$$

This velocity (310 m/s) is equal to 1,116 km/h. This is how fast the rocket would have moved in the absence of air resistance. In the actual launch, the velocity after 2 s was 222 m/s, or 800 km/h. Therefore, air resistance resulted in the velocity being about 72% of what it would have been.

Rocket Propulsion

Engineers have developed several ways of propelling large rockets. They use either solid fuel or liquid fuel for rockets designed to launch a satellite into orbit. In the case of smaller rockets already in outer space, they use pressurized gas or ion propulsion. It may also be possible to use a nuclear reactor someday as a source of propulsion.

Satellites

In this section, we will describe some major steps in the construction and launch of a satellite. Afterwards, we will describe the different satellite orbits and how instruments are placed in orbits around other bodies. Finally, the Lagrangian points, satellite power sources and the rendezvous of two objects in orbit will be discussed.

Construction and Launch of a Satellite

Once technicians assemble a satellite it is subjected to several tests. This is to insure proper operation in outer space. It may undergo a spin test to ensure that it survives being spun in space. It should undergo the vibration test. This is because the ride on a rocket is often bumpy. If instruments do not function after the vibration test, changes would have to be made. In many cases an electromagnetic test is also carried out. The purpose of this is to insure that one instrument's electric and magnetic field does not affect another instrument. A satellite should also pass a thermal vacuum test. Essentially, it is placed into a chamber and the air is pumped out to simulate the vacuum of space. During this test, the instruments must operate properly. Once a satellite has passed these tests, it is ready for launch.

A satellite launch is something that requires lots of planning. The appropriate launch vehicle is selected. A large and more expensive vehicle should not be used when a smaller one will do the job. A rocket must be launched over unpopulated areas.

The launch date is selected based on several factors, including the positions of current satellites. The Collision Avoidance List gives information about large objects orbiting Earth and is consulted. Once a launch time is selected, warnings

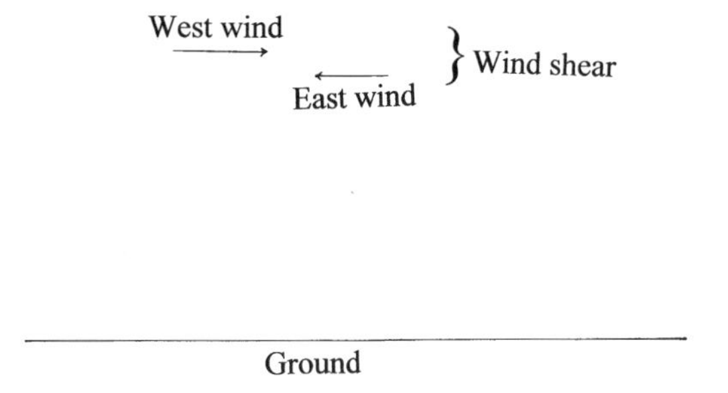

Fig. 1.13. An illustration of wind shear. Essentially, this occurs when two nearby layers of air move in different directions (Credit: Richard Schmude, Jr.).

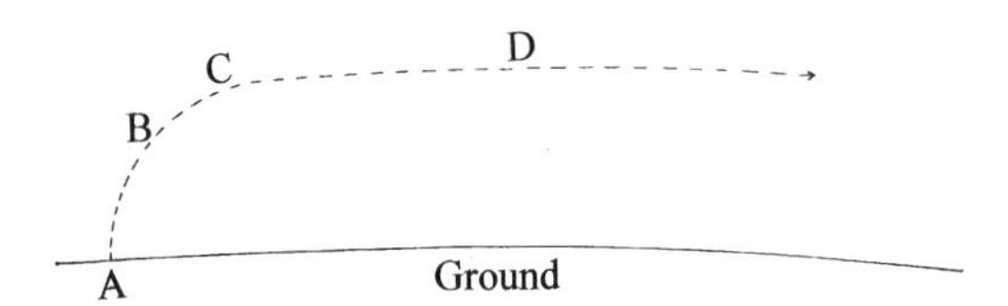

Fig. 1.14. The four basic steps in the launch of a satellite with a three-stage rocket; these steps are *A* the rocket lifts off; *B* the first stage lifts the rocket to a certain altitude, runs out of fuel and is ejected; *C* the second stage lifts the rocket higher, runs out of fuel and is ejected; and *D* the third stage lifts the satellite into orbit, runs out of fuel and is ejected (Credit: Richard Schmude, Jr.).

are issued to ships to avoid the areas below the rocket path. Shortly before launch, the fuel is put into the rocket.

Before launch, meteorologists are consulted for the latest weather updates. Wind shear can destroy a rocket and is illustrated in Fig. 1.13. It occurs when two nearby layers of air move in different directions. Other weather problems include lightning, high winds and strong air turbulence. Engineers and technicians also check the equipment. Once everything is ready, engines are fired and the rocket lifts off.

Figure 1.14 illustrates the basic steps in a typical launch. The rocket blasts off and moves straight up. As it moves, it accelerates, and as a result its velocity increases. As it gains altitude, engines are adjusted, causing the trajectory to bend. Smaller engines are adjusted to compensate for any shifts in the trajectory. About 2 min after liftoff, the rocket attains an altitude of about 50 km. At this point it is moving at an angle of about 45°. This is also about the time when the first stage runs out of fuel and is ejected. The empty stage falls. (This is one reason why rockets are launched over unpopulated areas.) The second stage engine ignites. It continues to accelerate and changes direction until it is almost parallel to Earth's surface. Once the second stage runs out of fuel, it is ejected and also falls. The third stage fires its engine and accelerates until it moves at the required velocity. At this point the satellite may be 400 km above Earth's surface. When the third stage runs out of fuel it is ejected and the payload is released into orbit. In some cases a satellite has small engines. These allow engineers to adjust its orbit. Unlike the first two stages, the third stage may remain in orbit around Earth.

Once a satellite moves fast enough it orbits Earth. What is happening here is that it moves straight and falls to Earth at the same time. This is why the orbit is curved. See Fig. 1.15. We see this same curved path every time one throws a ball. Essentially, the ball moves straight and falls at the same time. See Fig. 1.16. When it is thrown harder it moves father but still follows a curved path.

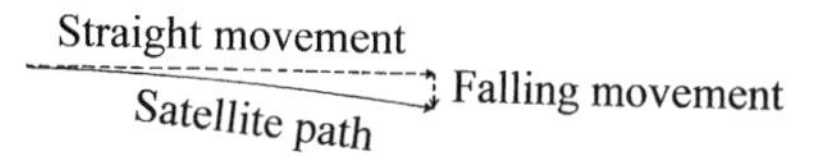

Fig. 1.15. A satellite undergoes two movements simultaneously, namely straight and downward. These are shown as *dashed lines*. The result of these two movements is the curved path (*solid line*) (Credit: Richard Schmude, Jr.).

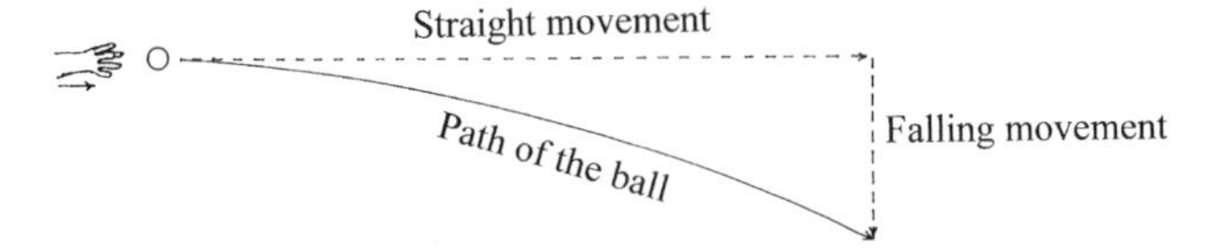

Fig. 1.16. A ball is thrown straight and it undergoes two simultaneous movements. These are in the direction thrown and downward. The two movements are shown as *dashed lines*. The result of these two movements is the *solid curved path* (Credit: Richard Schmude, Jr.).

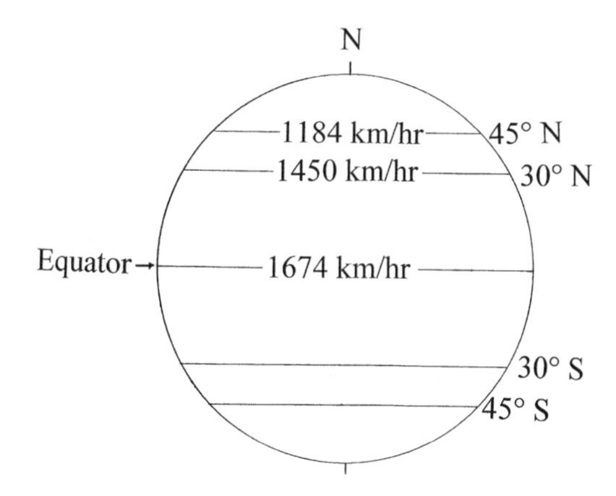

Fig. 1.17. Rotational speeds in km/h for areas at the equator, 30° and 45°N/S (Credit: Richard Schmude, Jr.).

One of the great achievements of Sir Isaac Newton was that he proved that gravity causes an apple to fall from a tree and is what causes the Moon to move around Earth. He proved that Earth's gravitational field extends into space. The further one is from Earth, the weaker Earth's gravitational tug becomes. In fact, satellites orbiting Earth experience small gravitational tugs from both our Moon and Sun.

Did you know that Earth's movement can help launch a satellite into orbit? This is because of rotation. Since the equator is wider than other latitudes, it has the highest rotational speed. See Fig. 1.17. The speed of the Earth at the equator is 1,674 km/h. Since a rocket standing at the equator moves with Earth, it has an eastward speed of 1,674 km/h with respect to space. On the other hand, at 45°N latitude the corresponding velocity is 1,184 km/h.

Space agencies want launch sites to be near the equator. This is a major reason why the European Space Agency has a launch site in French Guiana (along the northeastern coast of South America). It is near the equator.

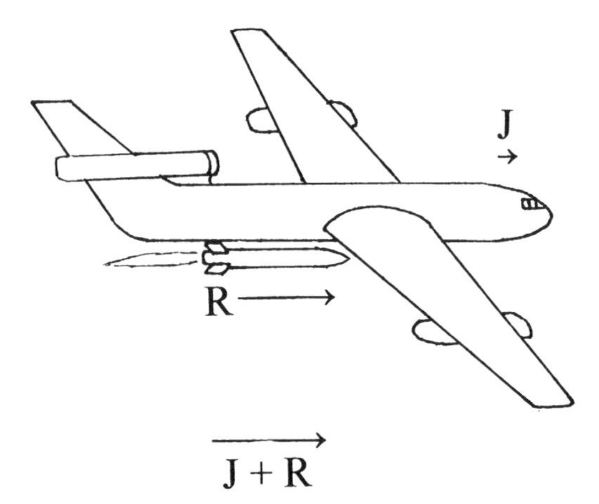

Fig. 1.18. The Interstellar Boundary Explorer (IBEX) satellite was launched with a Pegasus rocket from a high altitude jet plane. The speed of the high altitude jet plane with respect to outer space (a sum of its speed and Earth's rotational speed) is *J*, the speed of the rocket with respect to the jet after launch is *R* and the speed of the rocket with respect to outer space is *J*+*R*. The two advantages of this launch are the rocket gets an extra boost from the moving plane, and the launch is several kilometers above sea level and, hence, the satellite does not have to travel as far to get to outer space (Credit: Richard Schmude, Jr.).

One may get an additional boost if a jet-powered aircraft is used as a launch site. Engineers launched the Interstellar Boundary Explorer (IBEX) satellite with a Pegasus rocket that was attached to a high-flying jet aircraft. Since the aircraft was moving eastward it gave the satellite an extra boost. See Fig. 1.18.

Satellite Orbits

Satellites may follow either a circular orbit or an elliptical orbit. In a circular one, an object is always at the same distance from Earth's center. In an elliptical one, the distance between the satellite and Earth's center changes. The orbits are illustrated in Fig. 1.19. Perigee is the point closest to Earth and apogee is the point farthest from Earth.

The size of the orbit determines how fast a satellite moves. Table 1.4 lists orbital periods for several circular orbits. The altitude is the shortest distance between the satellite and Earth's surface.

Four common categories of orbits are low-Earth, low-eccentricity; low-Earth, high-eccentricity; high-Earth, high-eccentricity; and geostationary. All four are illustrated in Fig. 1.20. An example of a satellite flying in low-Earth, low eccentricity orbit is the International Space Station (ISS). On June 18, 2010, it was 350 km above Earth's surface. On this date, the orbital eccentricity was 0.02. The Russian *Molniya* satellite fits into the second category. This satellite serves the population at high northern latitudes. The Chandra X-Ray Observatory is an example of a satellite in the third category. It has a perigee distance of 29,000 km and an apogee distance of 120,000 km. *Amazonas 2,* a communications satellite, follows a geostationary orbit. This satellite's orbital period equals Earth's rotational period. Therefore, its position relative to Earth does not change much.

The orbital inclination is the angle between the satellite's orbital plane and Earth's equator. When the inclination equals 0° the satellite moves overhead at the

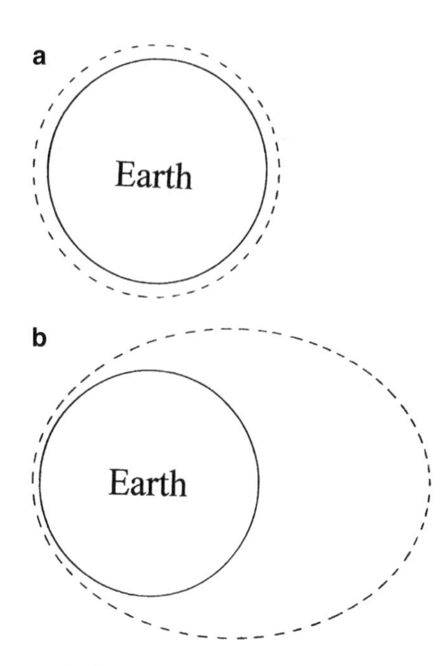

Fig. 1.19. A circular (**a**) and elliptical (**b**) orbit. In both drawings, the *dashed curve* is the path of the satellite (Credit: Richard Schmude, Jr.).

Table 1.4. Relationship between altitude and orbital period for satellites in circular orbits around Earth. The author computed all values

Altitude[a] (Km)	Orbital period (h)	Satellite to Earth's center distance (km)
250	1.48	6,628
500	1.57	6,878
750	1.66	7,128
1,000	1.74	7,378
35,786	24	42,164[b]

[a] Assuming that Earth's surface is 6,378 km from Earth's center
[b] Average distance of several geostationary satellites

equator. When the inclination equals 90° the satellite moves over the poles. Many satellites that can be observed have inclinations close to 90°. Geostationary satellites have inclinations near 0°.

Does a satellite fly over the same location on Earth? With the exception of geostationary satellites, the answer is no. This is because Earth is always rotating. The ground trace is the track that an orbiting object makes over Earth's surface. For example, let's assume a satellite is launched from Hammaguira French Special Weapons Test Centre in Algeria in a polar orbit. The orbital period is 1.5 h. Although the satellite is moving fast, Earth rotates at a rate of 15° per hour, or 22.5° for each orbit. As a result, the ground trace shifts 22.5° of longitude every 1.5 h. See Fig. 1.21.

A space station is an orbiting artificial satellite that is able to support one or more astronauts for an extended time. Table 1.5 lists a few space stations. Two of these lasted for over a decade. The two largest, *Mir* and the ISS had to be assembled

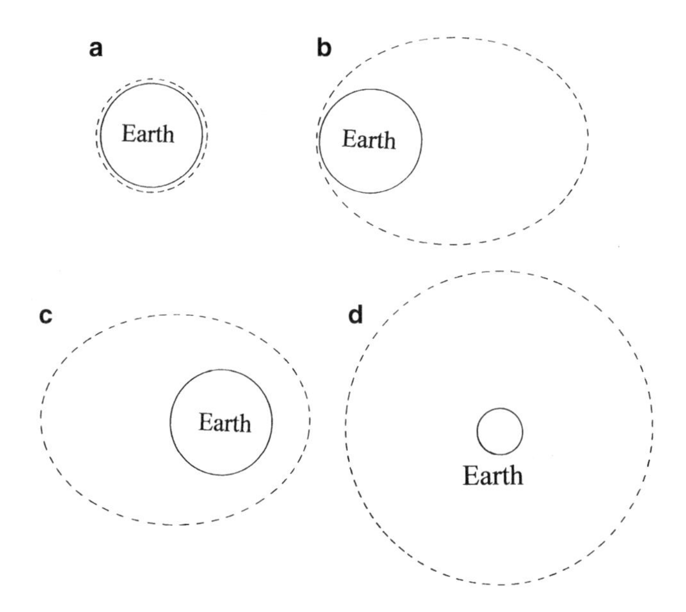

Fig. 1.20. Four types of satellite orbits (**a**) low-Earth orbit, low-eccentricity; (**b**) low-Earth orbit, high-eccentricity; (**c**) high-Earth orbit, high-eccentricity; and (**d**) geostationary (Credit: Richard Schmude, Jr.).

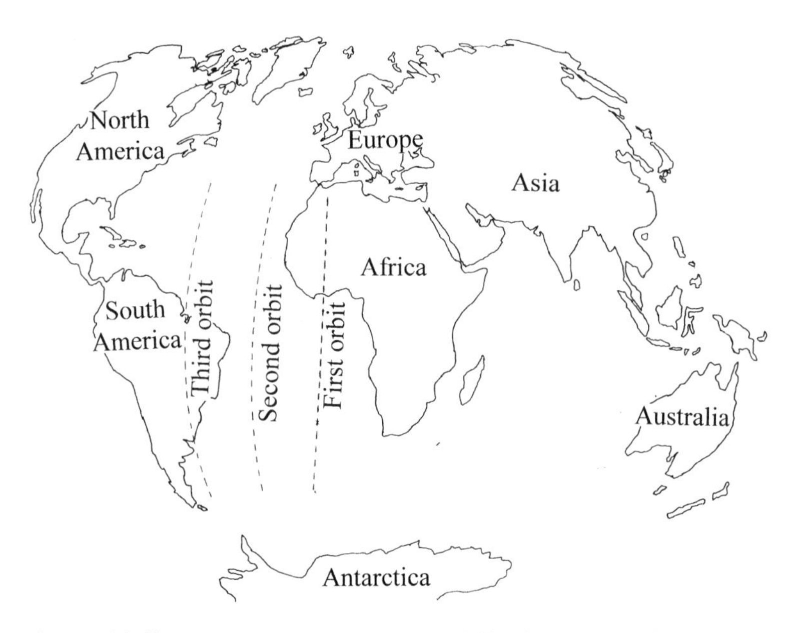

Fig. 1.21. The *dashed lines* show part of the ground trace of a satellite launched from the Hammaguira French Special Weapons Test Centre in Algeria in a 1.5 h polar orbit. When it makes its second orbit, its ground trace will be 22.5° of longitude farther west than during its first orbit. Likewise the ground trace of the third orbit will be 22.5° farther west than what it was for the previous orbit. This shift occurs because Earth moves as the satellite orbits Earth (Credit: Richard Schmude, Jr.).

over a period of several years. The Chinese launched the first piece of its space station, *Tiangong-1,* in September, 2011. They plan to add modules to *Tiangong-1* in the second decade of the twenty-first century.

Table 1.5. A summary of past and present space stations

Name	Longest length (m)	Empty weight (N)	Typical crew	Years of operation	Average altitude (km)
Salyut	15[a]	320,000[a]	2–4[b]	1971–1986	266[c]
Skylab	36[d]	890,000[e, d]	3[e]	1973–1979[f]	439[g]
Mir	33[h]	1,280,000[i]	Several	1986–2001[g, h]	434[j]
ISS	109[k]	4,120,000[i]	7[l]	1998-present	400[k]
Tiangong-1	10.4[m]	85,000[i]	3[i]	2011-present	360[i]

[a]Koutchmy and Nikol'skij (1983): 23–25
[b]Lutskij (1982): 33–34
[c]Computed from a period of 89.7 min orbital period as listed in *Sky & Telescope* (1974) and an average Earth radius of 6,371 km
[d]Includes an Apollo module docked from Lewis (1990): 9
[e]Damon (1989): 163
[f]Damon (1989): 10 and 73
[g]Average initial altitude; see *Sky & Telescope* 46: 22
[h]Oberg (2008b)
[i]Wikipedia – online encyclopedia
[j]Lewis (1990): 47
[k]Oberg (2008a)
[l]Janssen et al. (2011): 332
[m]First module only

Launching a Satellite Beyond Earth

Engineers have used several methods to launch satellites to other planets. One of these is to launch something fast enough to escape Earth's gravitational field. In essence, the rocket does all of the work of pushing the satellite to the point where it escapes. In the second method, the rocket does most of the work and the satellite does the rest. Essentially, the rocket pushes the satellite into an elliptical transfer orbit. Afterwards, engineers fire the satellite's engine and it escapes from Earth.

A third method of launching a satellite to another planet is the gravitational slingshot technique. Essentially, the rocket does some of the work and the gravitational field of a massive object does the rest. See Fig. 1.22. For example, engineers used the gravitational field of Jupiter to both boost the speed of the *New Horizons* spacecraft and to change its trajectory.

How do engineers cause a fast-moving satellite to orbit a planet? There are two ways of doing this. One of these is to use rockets to adjust the velocity. Essentially this causes the velocity to drop. This in turn will cause the planet to capture the satellite. A second way of doing this is aerobraking. This is a technique whereby one utilizes an atmosphere to slow down a spacecraft. Let's describe a recent example.

Engineers used aerobraking to insert the *Mars Reconnaissance Orbiter (MRO)* into the desired orbit around Mars. In this case, the *MRO* arrived at Mars in a highly elliptical orbit. Over the next few weeks, engineers forced the *MRO* to dip to an altitude of 105 km at perigee. The *MRO* was still traveling too fast to fall into its desired orbit at that time. See Fig. 1.23a. When the *MRO* reached perigee Mars' atmosphere was thick enough to exert a gentle frictional force, which over time

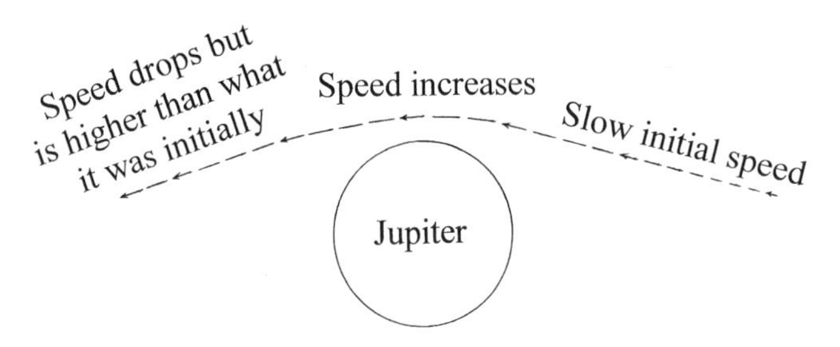

Fig. 1.22. In the gravitational slingshot method, a slow-moving spacecraft approaches Jupiter from the right, the spacecraft accelerates as it approaches Jupiter and its *speed increases.* As the spacecraft moves away it slows down, but the spacecraft is still moving faster than when it approached Jupiter. The length of the dash is proportional to the spacecraft speed (Credit: Richard Schmude, Jr.).

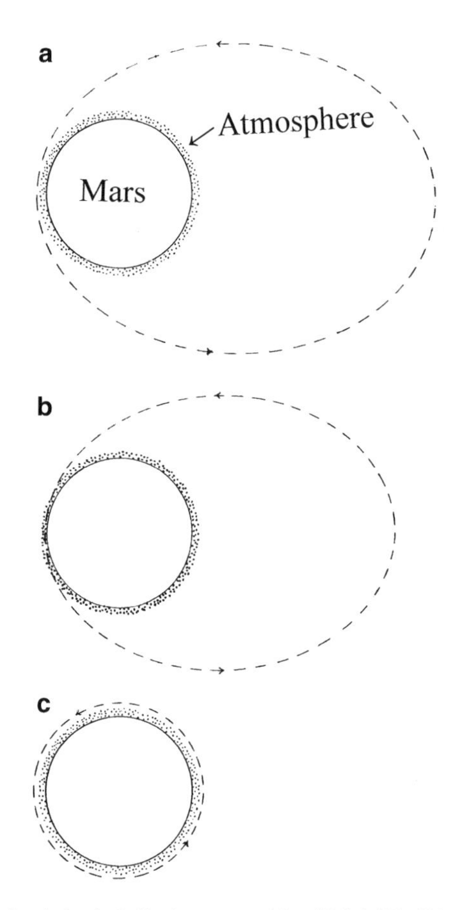

Fig. 1.23. The sequence of aerobraking for the Mars Reconnaissance Orbiter (MRO). (**a**) The MRO arrives at Mars in an elliptical orbit; (**b**) the altitude at closest approach is lowered until the MRO moves through Mars' upper atmosphere. Over a period of several weeks, the orbit becomes more circular. (**c**) The final orbit of the MRO (Credit: Richard Schmude, Jr.).

caused this satellite to slow down. See Fig. 1.23b. After several weeks of aerobraking, the *MRO* fell into the desired orbit. See Fig. 1.23c. The aerobraking technique is like a driver gently tapping his or her brakes. A rocket burn on the other hand is like a driver slamming on his or her brakes.

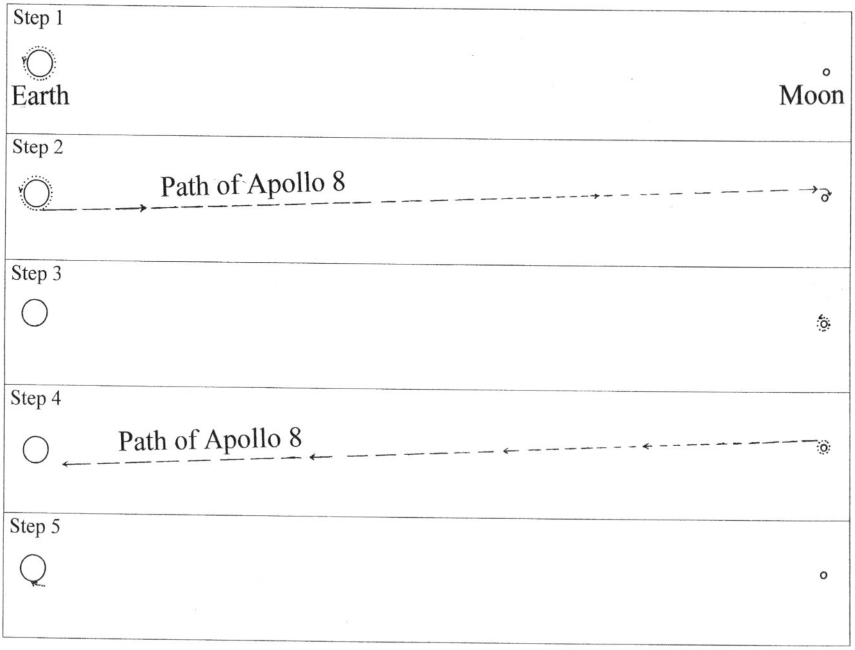

Fig. 1.24. Major steps in the *Apollo 8* mission. *Step 1* – Rocket launches the Command Service Module (CSM) into low-Earth orbit; *Step 2* – Astronauts fire a rocket, the CSM escapes Earth's gravity and moves to the Moon; *Step 3* – Astronauts reduce their speed with a rocket burn and the CSM orbits the Moon; *Step 4* – After ten orbits, astronauts fire a rocket and the CSM escapes from the *Moon* and heads back to Earth; and *Step 5* – Astronauts cause the CSM to slow down with a rocket burn and it falls back to Earth. In *Steps 2* and *4*, the length of the dash is proportional to the CSM speed (Credit: Richard Schmude, Jr.).

Sending a satellite to another body such as the Moon requires several steps. To illustrate this, let's recount humankind's first journey to the Moon.

Figure 1.24 summarizes briefly the *Apollo 8* Moon mission. This is based on the description in Baker (1981). In this mission, three astronauts (Bill Anders, Frank Borman and Jim Lovell) blasted off from Earth and entered a low-Earth, low-eccentricity orbit. They followed this path for about 2.5 h before firing the third-stage rocket. This enabled them to escape form Earth's gravity (a first for humankind). A short time later, the third stage was ejected. At this point, the CSM (Command Service Module) was moving at a speed of almost 39,000 km/h.

Earth's gravity caused the CSM to slow down as it moved away from Earth. As a result, its speed fell to 4,300 km/h 2.4 days later. At this point, however, it entered the Moon's gravitational field. This was an historical event. As the CSM approached the Moon, its speed increased to 9,400 km/h. This was too high for the spacecraft to orbit, and, hence, the astronauts fired a rocket engine. This reduced the speed.

The CSM orbited the Moon, and after ten orbits, the crew fired the rocket, causing it to escape. As it moved farther from the Moon, its speed fell. When the spacecraft crossed the point where it was under the influence of Earth's gravity its speed increased. It fell back to the home planet. Once the crew was close enough to Earth, they fired the rocket one more time, causing the CSM to slow down and fall back safely through the atmosphere.

Table 1.6. Kepler's third law for bodies in our solar system. The author computed all values listed below

Body	Mass (kg)[a, b]	k (day^2/km^3)
Sun	1.99×10^{30}	3.98×10^{-20}
Mercury	3.30×10^{23}	2.38×10^{-13}
Venus	4.87×10^{24}	1.61×10^{-14}
Earth	5.97×10^{24}	1.31×10^{-14}
Mars	6.42×10^{23}	1.23×10^{-13}
Jupiter	1.90×10^{27}	4.17×10^{-17}
Saturn	5.69×10^{26}	1.39×10^{-16}
Earth's Moon	7.35×10^{22}	1.07×10^{-12}
Ceres[c, d]	1.1×10^{21}	7.0×10^{-11}
Asteroid 4 (Vesta)[e, d]	1.8×10^{20}	4.5×10^{-10}
Asteroid 433 (Eros)[f, g]	6.69×10^{15}	1.2×10^{-5}

[a] Mass of the Sun is from McFadden et al. (2007)
[b] Mass of the planets and Moon are from *Astronomical Almanac* for the year 2010 (2008): E4
[c] Mass computed from a mean diameter of 952 km, a spherical geometry and an assumed average density of 2.5 g/cm^3
[d] Mean diameters of Vesta and Ceres are from the *Astronomical Almanac* for the Year 2010 (2008): G2
[e] Mass computed from a mean diameter of 512 km, a spherical geometry and an assumed average density of 2.5 g/cm^3
[f] Mass is from Souchay et al. (2003), who cites Miller et al. (2002)
[g] Value of k is approximate because of Eros' irregular shape

It takes a satellite less time to orbit our Moon than Earth. This is because our Moon's gravity is weaker. Kepler's Third Law predicts the orbital period around any particular body using:

$$P = \left(a^3 k\right)^{1/2} \tag{1.5}$$

In this equation, P is the orbital period in days, a is the average distance to a body's center in km and k is a constant. Values of k for different Solar System bodies are listed in Table 1.6. Table 1.7 lists the first satellites to orbit other Solar System bodies.

Lagrangian Points

In 1772, Joseph Lagrange demonstrated that in the presence of two high-mass bodies, m_1 and m_2, a third body, m_3, may remain in a stable position. Five areas where this occurs are the Lagrangian points and are labeled L_1, L_2, L_3, L_4 and L_5. The positions of L_1, L_2, L_4 and L_5 are shown in Fig. 1.25 for Earth and the Sun. The L_3 point lies on the Earth-Sun line and is opposite from Earth. The L_1, L_2 and L_3 points cover small areas. If a satellite is at one of these points and experiences a small force, it will begin to oscillate. This is not the case, however, if it is at L_4 or L_5. This is because L_4 and L_5 are larger areas of stability. In order for L_4 and L_5 to remain as areas of stability, m_2 (Earth) must be less than 0.04 times the mass of m_1 (Sun) and m_3 (satellite) must have a lower mass than m_2. Several satellites are at the Earth-Sun Lagrangian points.

Table 1.7. Successful satellites orbiting Earth and other celestial bodies

Satellite (country)	Object	Launch date	Date of orbital insertion
Sputnik (USSR) *First successful satellite to orbit the Earth*	Earth	Oct. 4, 1957[a]	Oct. 4, 1957
Vostok 1 (USSR) *First man to orbit the Earth*	Earth	April 12, 1961[b]	April 12, 1961[a,c]
Luna 10 (USSR) *First successful satellite to orbit the Moon*	Moon	March 31, 1966[d]	April, 1966
Apollo 8 (USA) *First men to orbit the Moon*	Moon	Dec. 21, 1968[d]	Dec. 24, 1968
Mariner IX (USA) *First successful satellite to orbit Mars*	Mars	May 30, 1971[d]	Nov. 13, 1971[d]
Venura 9 (USSR) *First successful satellite to orbit Venus*	Venus	June 8, 1975[d]	Oct. 22, 1975[d]
Galileo (USA) *First successful satellite to orbit Jupiter*	Jupiter	Oct. 18, 1989[d]	Dec. 7, 1995[d]
Cassini (USA) *First successful satellite to orbit Saturn*	Saturn	Oct. 15, 1997[d]	June 30, 2004[d]
Near-Earth Asteroid Rendezvous (USA) *First successful satellite to orbit an asteroid*	Eros	Feb. 17, 1996[d]	Feb. 14, 2000[d]
MESSENGER (USA) *First successful satellite to orbit Mercury*	Mercury	Aug. 3, 2004	March 18, 2011[d]
Dawn (USA) *First successful satellite to orbit Vesta*	Vesta[e]	June 2007	July, 2011

[a] Damon (1989): 3
[b] Zimmerman (2000)
[c] Peterson et al. (1998): 7
[d] McFadden et al. (2007): 904–909; Angelo (2000): 260
[e] Dawn will first orbit Vesta in 2011 and after its mission is complete, it will orbit Ceres

Satellite Power Sources

In order for a satellite to function properly, it needs power. Power is needed to keep the computer, communications equipment, scientific instruments and thrusters functioning properly. Three sources of power are solar panels, batteries and radioisotope thermoelectric generators (RTGs). Each of these will be described.

Solar panels are an ideal way for satellites to generate power. Essentially, sunlight strikes a panel of solar cells that convert the sunlight into electricity. Solar power is often used to recharge batteries. In the vicinity of Earth, the Sun gives off 1,366 W/m^2. This means sunlight having an energy of 1,366 J strikes a 1 m^2 surface each second. The surface must face the Sun directly. See Fig. 1.26. The efficiency is the fraction of sunlight power that is converted into useful forms.

For example, a solar panel with an efficiency of 0.100 (or 10.0%) converts one-tenth of the sunlight into electricity. Therefore, when an Earth-orbiting satellite

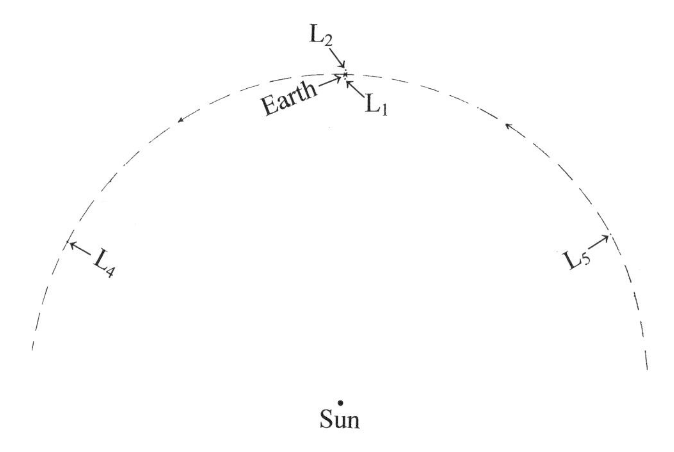

Fig. 1.25. Four of the five Lagrangian points for the Earth-Sun system are illustrated. The *L1* and *L2* points lie along the Earth-Sun line and are 1.5 million km from Earth. The *L4* and *L5* points are 150 million km from Earth and the Sun. The *L3* point lies on the opposite side of the Sun from Earth and is not shown (Credit: Richard Schmude, Jr.).

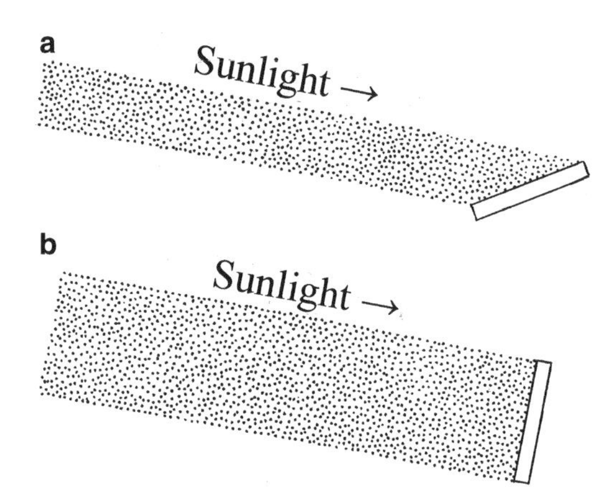

Fig. 1.26. When a surface is not facing the Sun directly, as is shown in *a*, it does not receive as much sunlight (*dotted area*) as the surface that faces the Sun directly *b* (Credit: Richard Schmude, Jr.).

faces our Sun directly, has 2.80 m^2 of solar panels, and has an efficiency of 0.100 it should produce 382 W. This is shown below.

$$\text{Total power} = 2.80\text{m}^2 \times 1366\ \text{W}/\text{m}^2 \times 0.100 = 382.48\ \text{W or } 382\ \text{W}. \quad (1.6)$$

Solar panels generate less power as they get further from our Sun. Therefore, satellites orbiting Solar System bodies beyond Mars usually have additional power sources besides solar panels.

Engineers have used radioisotope thermoelectric generators (RTGs) for a few satellites. An RTG converts heat, generated from the radioactive decay of an element, into electricity. The RTGs on *Cassini* contain plutonium-238 (^{238}Pu). Each RTG provided 295 W of power in 1997. The decay of ^{238}Pu caused the RTGs to produce less power during the first decade of the twenty-first century. At some point, the RTGs will no longer be able to supply enough power to keep *Cassini* functioning.

Batteries are a third power source. They contain two or more substances that react to produce an electrical potential or voltage. Different batteries produce different voltages. For example, the common AA battery produces an electrical potential of 1.5 V. Satellite batteries are rechargeable.

Rendezvous

How does one steer a satellite to a second one moving thousands of miles per hour? The answer is, very carefully. Essentially, rendezvous is the task of steering one satellite toward a second one. The two objects should have the same orbit and nearly the same velocity when they meet. There are several examples of rendezvous. For example, every time a cargo ship docked with the ISS, it underwent rendezvous. It also occurred during the *Apollo* Moon landings. Finally, the Japanese spacecraft *Hayabusa* made a rendezvous with the Earth-crossing asteroid 25143 Itokawa in 2005. This mission is described further in Chap. 2.

Electromagnetic Radiation and the Doppler Effect

Radio waves, infrared radiation, visible light and gamma rays are all examples of electromagnetic radiation. All of these have both an electric and a magnetic wave. The waves are at 90° angles from each other. All electromagnetic waves travel at the speed of light (2.998×10^8 m/s) in a vacuum. The wavelength is the length of one wave. See Fig. 1.27.

The frequency is the number of waves that pass a fixed point each second. The product of the wavelength and frequency is the speed. Table 1.8 lists frequency ranges for different examples of electromagnetic radiation.

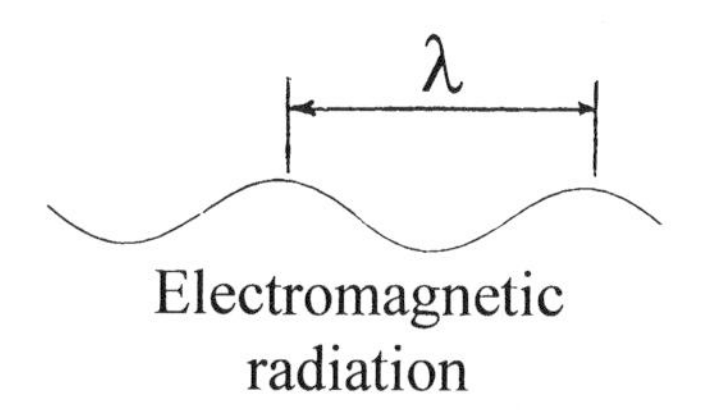

Fig. 1.27. The wavelength (λ) of electromagnetic radiation is the length of one complete cycle of the wave (Credit: Richard Schmude, Jr.).

Table 1.8. Different examples of electromagnetic radiation

Type	Wavelength range (m)	Frequency range (Hz)
Gamma[a]	1.2×10^{-13} to 1.2×10^{-10}	2.5×10^{18} to 2.5×10^{21}
X-rays[b]	1.0×10^{-11} to 1.0×10^{-8}	3.0×10^{16} to 3.0×10^{19}
Ultraviolet[c, d]	1.0×10^{-8} to 4.1×10^{-7}	7.3×10^{14} to 3.0×10^{16}
Visible[d]	4.1×10^{-7} to 7.1×10^{-7}	4.2×10^{14} to 7.3×10^{14}
Infrared[d, e]	7.1×10^{-7} to 1.0×10^{-3}	3.0×10^{11} to 4.2×10^{14}
Radio[f]	Greater than 1.0×10^{-3}	Below 3.0×10^{11}

[a] Angelo (2006): 257
[b] Angelo (2006): 657
[c] Angelo (2006): 622
[d] The author's observations with a calibrated spectroscope
[e] Angelo (2006): 312
[f] Angelo (2006): 487

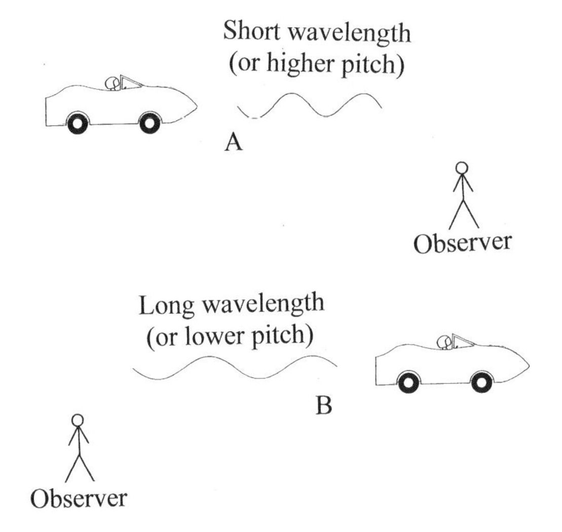

Fig. 1.28. As a race car approaches *A* its sound has a higher pitch (or shorter wavelength) than when it is moving away *B*. This is an example of the Doppler effect (Credit: Richard Schmude, Jr.).

Have you ever noticed how the sound of a race car changes as it approaches and passes? The sound has a higher pitch as it approaches and a lower one as it moves away. This is an example of the Doppler effect. Essentially, the frequency (or pitch) gets higher as the sound source (race car) approaches. As the car moves away, the pitch gets lower. See Fig. 1.28.

Electromagnetic radiation may also undergo a change in frequency. This occurs when the radiation source is moving towards or away from us. See Fig. 1.29. In this figure, the satellite gives off radio waves with a frequency of 9.00000 GHz. Since it is moving towards the receiver at a speed of 6 km/s, astronomers detect a frequency of 9.00018 GHz. They measure the frequency change and with this compute the satellite speed.

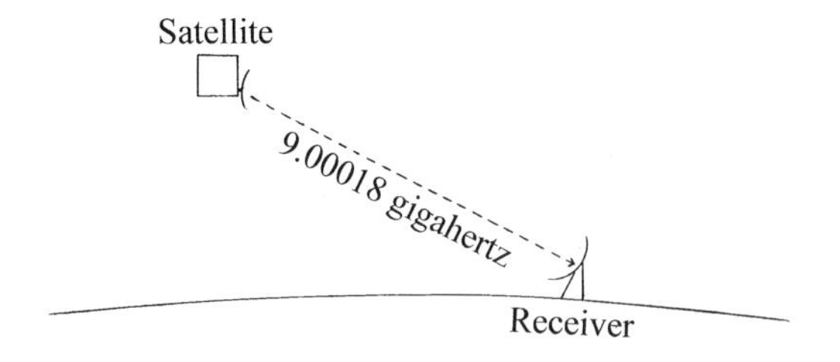

Fig. 1.29. A satellite uses a frequency of 9.00000 GHz. Since it is moving towards the receiver at a speed of 6 km/s, the frequency reaching it is 9.00018 GHz because of the Doppler effect (Credit: Richard Schmude, Jr.).

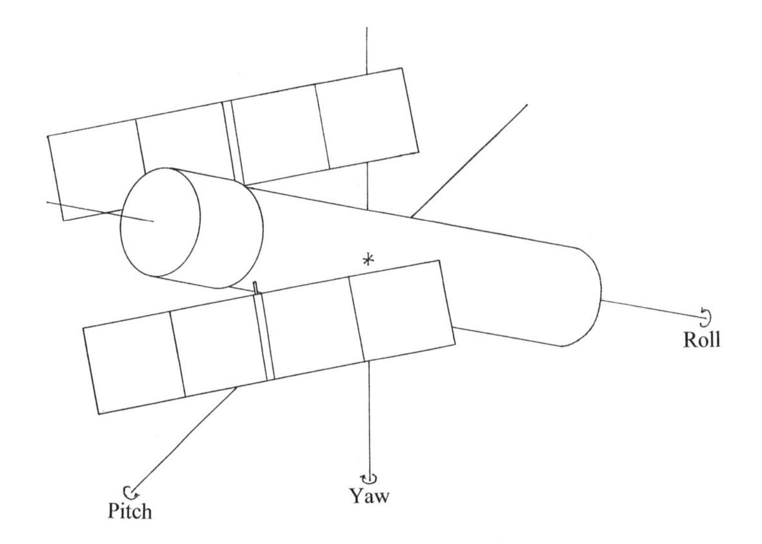

Fig. 1.30. The three dimensions of a satellite's position are *roll, pitch* and *yaw* (Credit: Richard Schmude, Jr.).

Satellite Communications and Tracking

Communication is carried out with radio waves. It is vital for keeping the satellite in the correct position, receiving data, sending commands and diagnostics.

Gyroscopes allow satellites to maintain the correct position, which is defined in three dimensions. See Fig. 1.30. These three are called roll, pitch and yaw. It is essential for a satellite to be in the correct position in order for it to transmit and receive commands. Furthermore, instruments must be pointed at the correct target.

Satellites are designed to transmit or relay radio signals back to Earth. Modest-sized radio dishes may detect nearby signals. Larger ones are needed for fainter signals coming from distant satellites. This is because signal intensity drops with increasing distance. The Deep Space Network is an array of large radio dishes in Goldstone, California; Canberra, Australia; and near Madrid, Spain. See Fig. 1.31. Several distant satellites were operating in late 2011; therefore, competition for facilities is great. China also has an extensive deep space network. Several radio receivers are spread across that country. See Fig. 1.32.

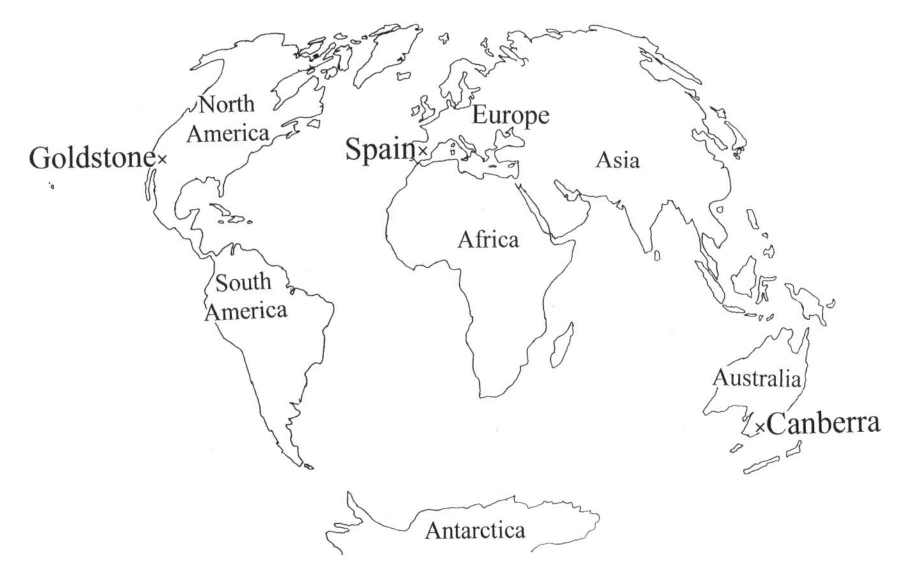

Fig. 1.31. The large radio dishes in the Deep Space Network are located in Goldstone, California; Canberra, Australia and near Madrid, Spain (Credit: Richard Schmude, Jr.).

Fig. 1.32. Large radio dishes in China are located in six different areas spanning almost 50° of longitude. The radio dishes are up to 50 m across. Locations are from McDowell (2006e) (Credit: Richard Schmude, Jr.).

Satellite Lifetime

Several factors affect a satellite's lifetime. For example, our Sun may cause our atmosphere to expand. This would lead to a satellite slowing down and burning up in the atmosphere. See Fig. 1.33. A collision with a small object may destroy instruments.

According to a NASA website, one of the *Themis* satellites was struck by an object in October 2010. Intense radiation from our Sun may cause a computer to shut

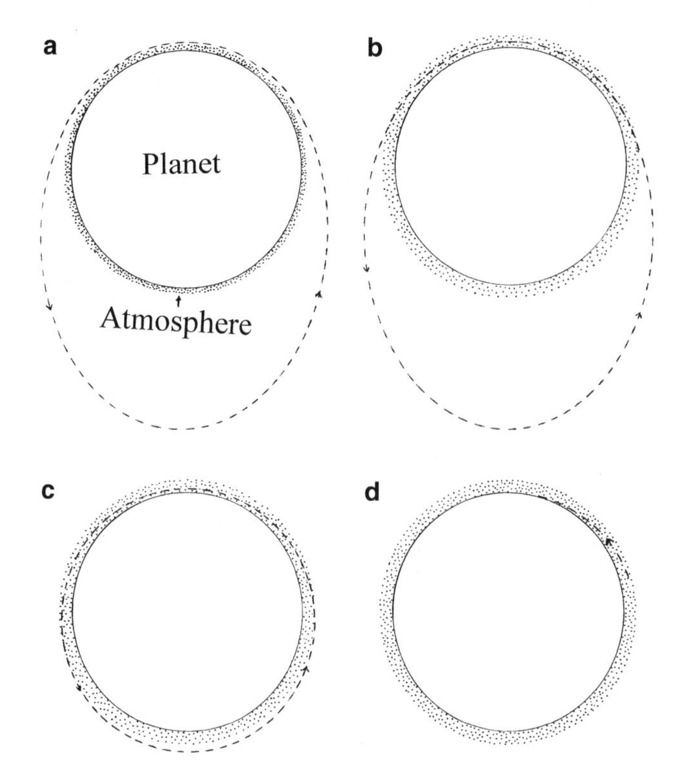

Fig. 1.33. The Sun may cause a planetary atmosphere to expand, which in turn may be disastrous for a satellite. (**a**) A satellite moves in an elliptical orbit and at its closest point is still above the atmosphere; (**b**) The atmosphere expands and, as a result, the satellite experiences a drag force as it moves through the upper atmosphere; (**c**) The orbit of the satellite changes over many weeks because of the new drag force; and (**d**) The satellite gets low enough for it to burn up (Credit: Richard Schmude, Jr.).

down. For example, McDowell (2009g) reports that the *Gravity Probe B* satellite temporarily shut down because of an intense burst of solar wind. If the thrusters fail, communication would become difficult. One reason for the Hubble Space Telescope servicing missions was to replace worn out gyroscopes. Without gyroscopes technicians cannot control its position. A loss of coolant may render an instrument useless. Many instruments require a coolant, such as liquid nitrogen, to function properly.

Several satellites also survive beyond their planned lifetime. In many cases, extra money is allocated. This allows astronomers to collect additional data. Table 1.9 lists examples of planned and extended lifetimes.

Satellite Visibility

Satellites reflect light from the Sun. This is why we see them. The amount of light they reflect depends on several factors, including albedo, phase, distance and size. It also depends on the mode of reflection (specular or diffuse).

The albedo is the fraction of light an object reflects. Freshly fallen snow reflects about 80% of the light falling on it, whereas charcoal reflects less than 10% of the light falling on it. The higher a satellite's albedo, the more light it reflects.

Table 1.9. A few satellites with the primary and extended mission lifetimes

Satellite (country/agency)	Target	Primary lifetime	Extended lifetime
Ulysses (ESA/USA)	Sun	1994–1995[a]	1995–2009[a]
Mars Global Surveyor (USA)	Mars	1999–2001[b]	2001–2006
Cassini (USA)	Saturn	2004–2008[c]	2008-present (2017)[d]
Mars Odyssey (USA)	Mars	2002–2004[e]	2004-present (2011)
Mars Exploration Rover Mission (USA)	Mars	January–April 2004[f]	2004-present (2011)
MIR (USSR/Russia)	Earth	1989–1994[g]	1994–2001
SOHO (ESA/USA)	Sun	1996–1998[h]	1998-present (2011)
Chang'e-1 (China)	Moon	1 year (2007–2008)	Four months beyond primary mission

[a] Angelo (2006): 260
[b] Angelo (2006): 383
[c] McDowell (2008j)
[d] *Sky & Telescope* (2010b)
[e] Angelo (2006): 381–382
[f] Angelo (2006): 382–383
[g] Peterson et al. (1998): 35
[h] Angelo (2006): 549

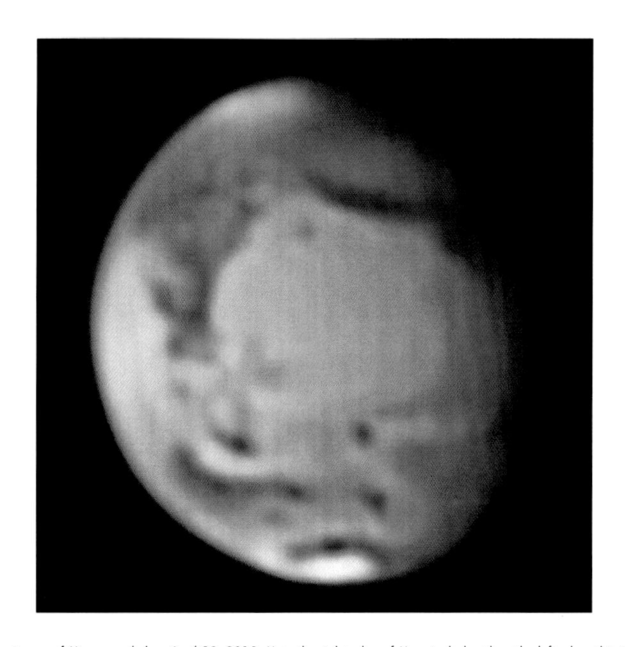

Fig. 1.34. An image of Mars recorded on April 28, 2010. Note the right edge of Mars is darker than the left edge; this is the combined result of millions of larger shadows on the right edge (Credit: Don Parker).

Which is brighter: a full Moon or a crescent Moon? The full moon is brighter. This is because: (1) it has a larger illuminated area and (2) it has almost no shadows. Shadows cause less light to be reflected and, hence, darken an area. See Fig. 1.34. Note that the right edge of Mars is darker than the left edge. This is caused by numerous shadows on the right edge. Most of the shadows are too small to see, but their combined effect is visible.

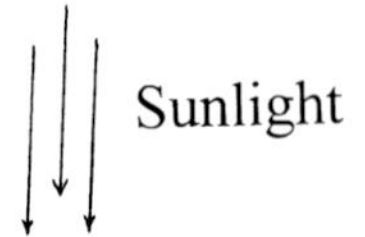

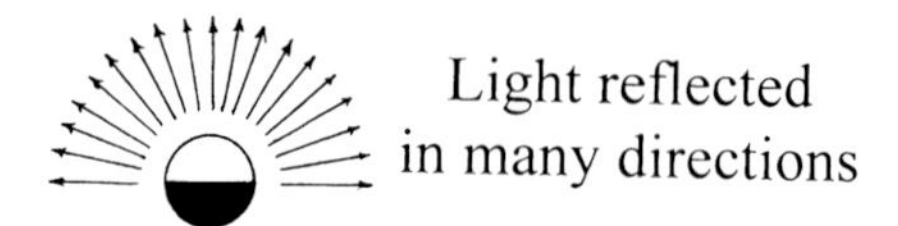

Fig. 1.35. A diffuse reflector reflects light in many directions, and as a result, the light intensity drops rapidly with distance (Credit: Richard Schmude, Jr.).

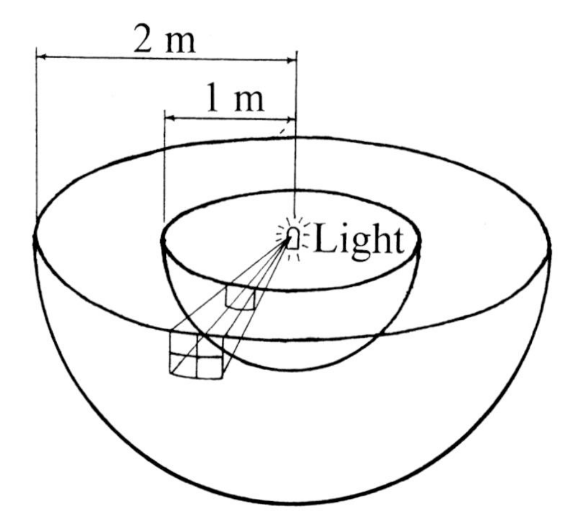

Fig. 1.36. The amount of light passing through one square at a distance of 1 m equals the amount of light passing through four equal squares at a distance of 2 m. This is why a lamp is four times brighter at a distance of 1 m than at 2 m (Credit: Richard Schmude, Jr.).

Satellites are visible because they reflect light from our Sun. Diffuse and specular are two modes of reflection. In most cases, a surface reflects sunlight in many directions. This is diffuse reflection. Since light is reflected in many directions, it grows dimmer with increasing distance. Figure 1.35 illustrates what is happening. Essentially, a diffuse reflector reflects light in many directions. In many cases, it reflects light in all directions within 90° of the line defined by the incoming light. Furthermore, the light is spread out uniformly in all directions. When this is the case, we have the situation shown in Fig. 1.36.

Essentially, the amount of light going through one square at a distance of 1 m equals the amount of light going through four squares at a distance of 2 m. Keep in mind that the amount of light entering our eye pupils controls the perceived brightness. Therefore, only one-fourth of the light at 1 m enters our eye at 2 m. Essentially, the brightness drops with the square of the distance. For example, a

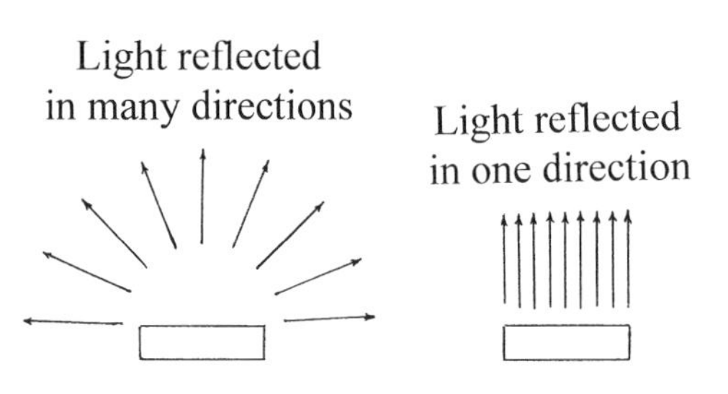

Fig. 1.37. Diffuse (*left*) and specular reflectors (*right*). The specular reflector will appear much brighter than the diffuse one. This is because the diffuse reflector spreads out the reflected light, but the specular reflector does not spread out the light (Credit: Richard Schmude, Jr.).

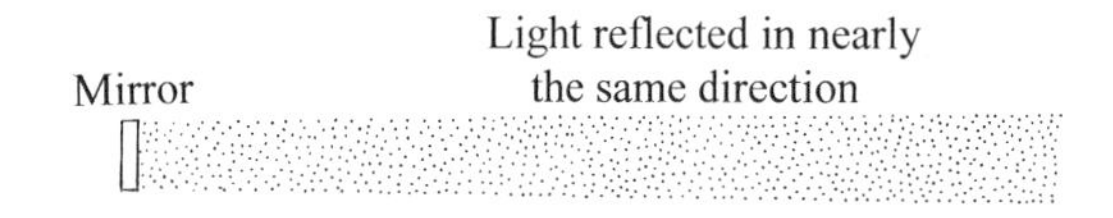

Fig. 1.38. The nearly specular reflection of the 11 cm mirror mentioned in the text is illustrated. Since the light spreads out a little, it grows a little dimmer with increasing distance (Credit: Richard Schmude, Jr.).

satellite 300 km away would appear four times brighter than if it is 600 km away. Therefore, as a satellite moves gracefully across the sky, its brightness will rise or fall because of its changing distance.

Specular reflection is different. Examine Fig. 1.37. Essentially, light is reflected in a specific direction instead of in many directions. One observes this mode of reflection with a mirror reflecting sunlight in a shaded area. Specular reflection is common. For example, the sand on a beach sparkles because the grains are oriented in such a way that light is reflected straight back to the observer. Most objects reflect light close to the diffuse or specular modes. An 11 cm mirror purchased for $2 reflects a bright patch of sunlight 27 m (90 ft) away. The bright patch has an area about six times larger than the mirror. Essentially, the light gradually spreads out at a small angle. See Fig. 1.38.

In the case of perfect specular reflection, the light is reflected back and the beam does not spread out. In the case of the mirror, the reflected beam of light spreads out gradually. The beam spreads because there are mirror imperfections and also because of scattering caused by air and suspended particles. Therefore, for a given amount of reflected light, the intensity does not drop with increasing distance for perfect specular reflection; it drops a little for nearly specular reflection and it drops rapidly for diffuse reflection.

Most satellites reflect diffusely. One exception is the *Iridium* satellite. It becomes bright because the mirror-like components are oriented just right. When this occurs, specular reflection takes place. *Iridium* satellites are discussed further in Chap. 4.

The size of a satellite also affects its brightness. A large satellite usually reflects more light than a smaller one. The cross-sectional areas of satellites range from over 1,000 m^2 for the ISS to less than 1.0 m^2 for smaller ones. If all other factors are equal, the ISS would be over 1,000 times brighter than the smaller satellite.

In many cases, a satellite may grow bright and dim periodically because it is tumbling and different cross-sectional areas are facing the observer. See Fig. 1.39.

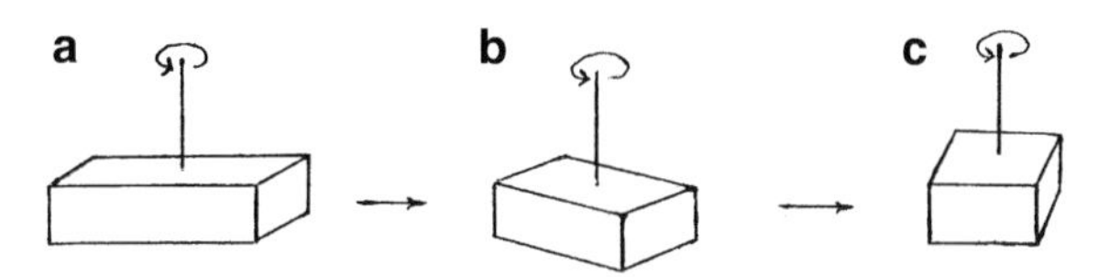

Fig. 1.39. As a satellite rotates, different sides face the viewer. If it is a diffuse reflector, it would be brightest at position a and dimmest at position c. This is because it has a larger cross-sectional area at position a than at c (Credit: Richard Schmude, Jr.).

Satellite Applications

Satellites today are used in a wide variety of ways. Some are designed to keep track of the weather. Weather images are frequently shown on television. Others are used for television, Sirius radio and other communications applications. Others monitor Earth's surface. Satellites are used for military applications, the Global Positioning System and for the study of other celestial bodies and the space environment. Satellites used in the study of objects beyond Earth are described in Chap. 2. Those that study Earth are described in Chap. 3.

Chapter 2

Scientific Satellite Spacecraft

Introduction

Many satellites are designed to gather information on extraterrestrial bodies. We will call these "spacecraft." One of the earliest spacecraft, *Luna 3,* imaged the far side of the Moon. The former USSR launched it on the second anniversary of *Sputnik 1.* Since 1959, the United States, the former USSR, Japan, China, India and the European Space Agency have launched spacecraft beyond Earth's orbit. The main objective of these has been to gather scientific data. Some have enabled astronomers to better understand our Sun and how it heats Earth, while others have given us a better understanding of the processes at work on other planets. In this chapter, we will give an overview of the different types of spacecraft missions. Afterwards, we will describe seven specific spacecraft in detail. These seven and the countries/agencies that launched them are: *Chang'e-1* (China), *Chandrayaan-1* (India), *Hayabusa* (Japan), Hubble Space Telescope (United States), *Cassini* (United States and the European Space Agency), *Spektr-R* (Russia) and *Gaia* (European Space Agency).

Spacecraft Missions

Table 2.1 lists the advantages and disadvantages of the flyby, rendezvous, orbiter, lander, sample return and human missions. Each of these will be described.

Figure 2.1 illustrates a flyby mission. Essentially, a spacecraft flies by its target and moves away. It does not orbit because it is moving too fast to be captured. In this mission, the spacecraft is close to its target for only a few days. For example, *Voyager 2* was only able to record useful images of the planet Uranus for about a week during its January 1986 flyby.

In many cases, a spacecraft will fly past a target of interest along the way to its destination. For example, *New Horizons* flew past Jupiter and its moons in 2007. It took close-up images and carried out measurements during this encounter. See Fig. 2.2.

Table 2.2 lists a few recent and planned flyby missions. In at least one case, an unplanned flyby took place. On January 23, 2000, Cassini happened to pass asteroid 2685 (Masursky). It was able to image that object at low resolution.

R. Schmude, Jr., *Artificial Satellites and How to Observe Them*, Astronomers' Observing Guides,
DOI 10.1007/978-1-4614-3915-8_2, © Springer Science+Business Media New York 2012

Table 2.1. Advantages and disadvantages of the six different types of spacecraft missions

Mission	Advantage	Disadvantage
Flyby	Low cost, useful for large and small bodies	Data is collected for just a short period of time
Rendezvous	Data may be collected for a long period of time	Extra fuel is required, difficult to use for large bodies
Orbiter	Data may be collected for a long period of time	More expensive than a flyby mission
Lander	Close-up images may be obtained; measurements of the deep atmosphere may be carried out.	Usually more expensive than the orbiter mission
Sample return	A sample of the target may be analyzed on Earth with the latest equipment	Only a small sample may be obtained. It is often very expensive
Human	An astronaut may solve problems as they arise; an astronaut is able to do more than robots	There is the chance of an accident and loss of life; this is more expensive than the other missions

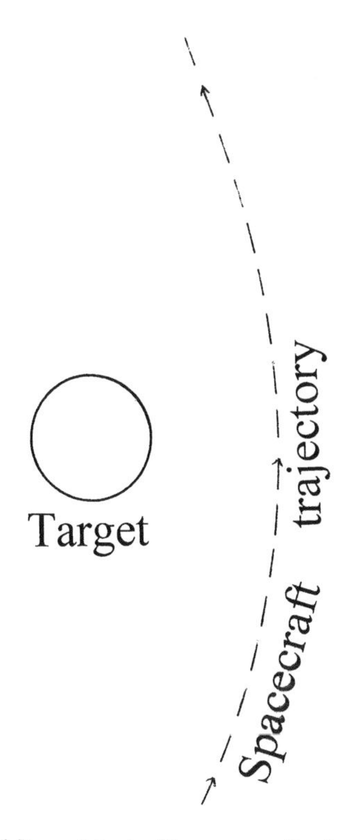

Fig. 2.1. In a flyby mission, a spacecraft flies near its target and then moves away. Note that the target's gravity alters the trajectory (Credit: Richard Schmude, Jr.).

In the rendezvous mission, a spacecraft approaches its target but does not orbit it. In this case, the target has such a weak gravitational pull that it would be difficult to adjust the velocity to initiate capture. This mission is illustrated in Fig. 2.3.

Japan's *Hayabusa* spacecraft carried out a rendezvous mission. It traveled alongside of asteroid 25143 Itokawa. It also released a small lander, which touched down on the surface of the asteroid. This soft landing also represents a rendezvous event.

Fig. 2.2. The long range reconnaissance imager (LORRI) on the *New Horizons* spacecraft took this image of Jupiter's moon Io on March 1, 2007 (Credit: NASA/Johns Hopkins University Applied Physics Laboratory/Southwest Research Institute).

Table 2.2. Recent and future flyby missions

Spacecraft	Entity	Target	Date	Source
Stardust	NASA	Asteroid 5535 (Annefrank)	November, 2002	McDowell (2007h)
Deep impact	NASA	Comet 9P/Temple 1	July 4, 2005	Schmude (2010)
Rosetta	ESA	Mars	February 2007	Tytell (2007)
New Horizons	NASA	Jupiter	February, 2007	Tytell (2007)
Messenger	NASA	Mercury	(2008–2009)	McDowell (2007g)
Stardust-NExT	NASA	Comet 9P/Temple 1	February 2011	McDowell (2007h)
Deep Impact (EPOXI)	NASA	Comet 103P/Hartley	November 2010	Sky and Telescope (2011a)
Rosetta	ESA	Asteroid 2867 (Steins)	2008	McDowell (2008i)
Rosetta	ESA	Asteroid 21 (Lutetia)	2010	Sierks et al. (2011)
New Horizons	NASA	Pluto	July 2015	New Horizons homepage see Table 2.10

The orbiter mission is used for a large target such as a planet. Essentially, a spacecraft orbits its target. This mission is usually more expensive than a flyby one for two reasons. Firstly, an orbiter requires a rocket with fuel. This is extra weight and, hence, more expensive to launch. Secondly, an orbiter mission lasts longer and, hence, costs are higher.

In a few cases, an orbiting spacecraft will carry additional fuel so that engineers may change its orbit. For example, the *Cassini* spacecraft made numerous close flybys of Saturn's moons. These maneuvers required extra rocket burns. Every time *Cassini* approached a moon, its orbit changed.

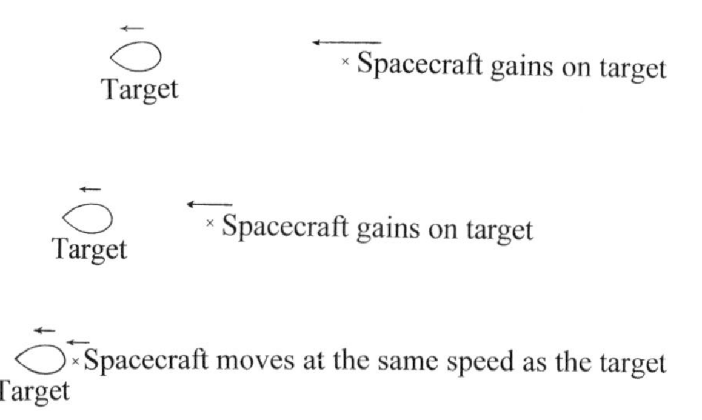

Fig. 2.3. In the rendezvous mission, a spacecraft approaches (or gains) on its target until it is close enough to make measurements. At this point, it moves with the target, as is shown at the bottom (Credit: Richard Schmude, Jr.).

A lander mission is one where a spacecraft studies the dense portions of a body's atmosphere and/or lands on its surface. Some spacecraft crash land. Others have a soft landing. A rover is a special kind of lander. It moves along the surface. Several lander missions are outlined in Table 2.3.

Since the 1990s, scientists have intentionally crashed objects into the Moon. In a number of cases, their goal was to detect water in the debris kicked up by the crashes. Table 2.4 lists a few deliberate lunar impacts. The objective of the *Lunar Crater Observation and Sensing Satellite (LCROSS)* mission was to crash two large objects into the Moon and image the debris cloud. The mission was successful. Beatty (2010a) reports that one group of astronomers detected significant amounts of water in the debris cloud.

The former USSR carried out the first robotic sample return mission, *Luna 16,* in September 1970. It landed in the Moon's Sea of Fertility and scooped up a sample of regolith, then blasted off back to Earth. The Soviet *Luna 20* (February 1972) and *Luna 24* (August 1976) spacecraft collected additional Moon samples. Since the 1970s, the Japanese and Americans have carried out successful sample return missions. Table 2.5 lists the successful sample return missions as of early 2011. The two American sample return missions (*Genesis* and *Stardust*) entailed spacecraft that collected particles in space. Neither spacecraft landed on its target. *Hayabusa* was different. A small lander from *Hayabusa* landed but malfunctioned. In spite of this, it collected dust near its target.

The manned mission entails considerable cost because a life support system is required. A human being needs a pressurized environment to survive. On Earth, the atmosphere exerts enough pressure to sustain human life. In space gas must be brought along. Astronauts also need air, food, water and various facilities. This material must be brought along.

The missions just described entail spacecraft that become satellites. These will orbit a planet, an asteroid, the Moon, the Sun or the galaxy center. Five spacecraft are heading out of our Solar System. These are *Pioneer 10* and *11, Voyager 1* and *2* and *New Horizons.* Eventually, they will probably end up orbiting the galaxy center.

Several dozen spacecraft were launched between 2000 and 2010. Some of these are designed to study one aspect of space and others are designed to study several objects. These can be divided into six categories based on their target: (1) the Sun, (2) the Moon, (3) Mars, (4) other Solar System objects besides the Sun, the Moon

Table 2.3. Selected lander missions

Body	Spacecraft (year)	Type (Entity)	Source
Moon	Ranger 7 (1964)	CL (NASA)	Watts (1968)
	Ranger 8 (1965)	CL (NASA)	"
	Ranger 9 (1965)	CL (NASA)	"
	Luna 9 (1966)	SL (USSR)	Zimmerman (2000) p. 47
	Surveyor 1 (1966)	SL (NASA)	Watts (1968)
	Luna 13 (1966)	SL (USSR)	Zimmerman (2000) p. 57
	Surveyor 3 (1967)	SL (NASA)	Watts (1968)
	Surveyor 5 (1967)	SL (NASA)	"
	Surveyor 6 (1967)	SL[a] (NASA)	"
	Surveyor 7 (1968)	SL (NASA)	"
	Luna 17 (1970)	R (USSR)	Zimmerman (2000) p. 90
	Luna 21 (1973)	R (USSR)	Zimmerman (2000) p. 112
	Luna 23 (1974)	SL (USSR)	Zimmerman (2000) p. 126
	Moon Impact Probe (2008)	CL (ISRO)	Chandrayaan-1 homepage (see Table 2.8)
Venus	Venera 4 (1967)	A (USSR)	Zimmerman (2000) p. 64
	Venera 5 (1969)	A (USSR)	Zimmerman (2000) p. 78
	Venera 6 (1969)	A (USSR)	Zimmerman (2000) p. 78
	Venera 7 (1970)	SL (USSR)	Sparrow (2009) p. 261
	Venera 8 (1972)	SL (USSR)	"
	Venera 9 (1975)	SL (USSR)	"
	Venera 10 (1975)	SL (USSR)	Zimmerman (2000) p. 136
	Venera 11 (1978)	SL (USSR)	Zimmerman (2000) p. 162
	Venera 12 (1978)	SL (USSR)	Zimmerman (2000) p. 161
	Venera 13 (1982)	SL (USSR)	Sparrow (2009) p. 261
	Venera 14 (1982)	SL (USSR)	"
Mars	Viking Lander 1 (1976)	SL (NASA)	Sparrow (2009) p. 262
	Viking Lander 2 (1976)	SL (NASA)	"
	Sojourner (1997)	R (NASA)	Sparrow (2009) p. 274
	Spirit/Opportunity (2004)	R (NASA)	McDowell (2009e)
	Phoenix (2008)	SL (NASA)	Beatty (2008)
Asteroid 433 (Eros)	NEAR Shoemaker (2001)	SL (NASA)	Beatty (2001)
Asteroid 25143 (Itokawa)	Hayabusa (2005)	SL (JAXA)	Beatty (2006a)
Jupiter	Galileo (1995)	A (NASA)	Sparrow (2009) p. 270
Titan	Huygens (2005)	SL (ESA)	Tytell (2005a)

A destroyed in the atmosphere, *CL* crash landed, *SL* soft landed, *R* rover

ISRO Indian Space Research Organization, *JAXA* Japanese Space Agency, *NASA* National Aeronautics and Space Administration, *USSR* Union of Soviet Socialist Republics

[a]Surveyor 6 lifted off the lunar surface and then touched down again

Table 2.4. Lunar impacts of human-made objects since 1999

Object	Date of impact	Location	Source
Lunar prospector	July 31, 1999	87.7°S, 42.1°E	Zimmerman (2000) p. 329
SMART-1	September 3, 2006	33.3°S, 46.2°W	Burchell et al. (2010)
MIP from Chandrayaan-1	November 14, 2008	Near the Moon's South Pole	Chandrayaan-1 homepage (See Table 2.8)
Chang'e-1	March 1, 2009	1.50°S, 52.36°W	McDowell (2009f)
Kaguya	June 11, 2009 (JST)	65.5°S, 80.4°E	Kaguya home page; see Table 2.8
LCROSS	October 9, 2009	84.7°S, 50°W	Redfern (2009) and Beatty (2010a)
Centaur rocket near LCROSS	October 9, 2009	84.7°S, 50°W	Redfern (2009) and Beatty (2010a)

Table 2.5. Successful sample return missions

Mission	Entity	Target	Date or year	Source
Luna 16	USSR	Lunar surface	September 1970	Sparrow (2009) p. 258
Luna 20	USSR	Lunar surface	February 1972	Sparrow (2009) p. 258
Luna 24	USSR	Lunar surface	August 1976	Sparrow (2009) p. 258
Stardust	NASA	Comet 81P/Wild	January 2004	Schmude (2010)
Genesis	NASA	Solar wind	2005	McDowell (2004)
Hayabusa	JAXA	Asteroid 25143 (Itokawa)	2005[a]	Sky and Telescope (2011b)

[a]We are not certain when the samples were collected because of equipment glitches

and Mars and (5) phenomena beyond our Solar System. A sixth category, space telescopes, study many targets. Characteristics and the home pages for these spacecraft are summarized in Tables 2.6, 2.7, 2.8, 2.9, 2.10, and 2.11.

Our atmosphere distorts images and is opaque to many wavelengths. Since the nineteenth century, astronomers have built observatories on mountains to minimize the degradation caused by our atmosphere. More recently, they have placed telescopes in outer space. These have produced high-resolution images. Furthermore, space telescopes detect X-ray, ultraviolet and far-infrared wavelengths that our atmosphere blocks. Therefore, they allow astronomers to use a variety of wavelengths in their studies. Figure 2.4 shows the frequencies at which a few of these telescopes operate, and Fig. 2.5 shows their mirror sizes.

Chang'e – 1

Chinese engineers launched their first Moon satellite, *Chang'e-1,* on October 24, 2007, using a Long March 3A (or Chang Zheng 3A) rocket. They launched it from Xiching spaceport. The satellite initially orbited Earth. It followed an elliptical-transfer orbit (ETO), which stretched out. With planned rocket burns, engineers were able to push it into the Moon's gravitational field. About 2 weeks after launch, *Chang'e – 1* was in a nearly circular orbit around the Moon. It was about 200 km above the lunar surface and orbited once every 2.1 h. Therefore, the Moon rotated 1.16° for each trip. This equals a shift of 35 km at the equator.

This mission lasted about 16 months. Once *Chang'e-1* completed its mission, it crash landed on the Moon.

The major scientific objectives of *Chang'e-1* and relevant instruments are summarized in Table 2.12. Figure 2.6 shows instruments and their frequency ranges. Most of the information in this section is from McDowell (2006e, 2008a), Huixian et al. (2005), and Zhi-Jian et al. (2005) and the website http://nssdc.gsfc.nasa.gov/nmc/spacecraftDisplay.do?id=2007-051A.

The *Chang'e – 1* orbiter had an initial mass of a large automobile. Solar arrays provided power for it. It contained a stereo camera, a spectrometer imager, a laser altimeter, a microwave radiometer, a particle detector, and spectrometers that detected X-rays and gamma rays. We will now describe these instruments.

Table 2.6. Space telescopes in operation as of 2011

Name or acronym	Year of launch	Entity	Orbit type (inclination)	Purpose	Homepage
HST	1990	NASA	LELE, (28.5°)	Survey the sky in ultraviolet, visible and infrared wavelengths	http://hubble.nasa.gov
Chandra	1999	NASA	HEHE	Obtain X-ray images and X-ray spectra of selected targets in the sky	http://Chandra.harvard.edu
FUSE	1999	NASA	LELE	Characterize stars, galaxies and interstellar gas in far ultraviolet light	http://fuse.pha.jhu.edu/
XMM-Newton	1999	NASA/ESA	HEHE (40°)	Observe high-energy X-rays from pulsars, black holes and active galaxies	http://xmm.sonoma.edu
GALEX	2003	NASA	LELE	Record ultraviolet images of galaxies, stars and quasars	http://www.galex.caltech.edu
Spitzer	2003	NASA	HET	Search for dust discs around stars, image the Milky Way Galaxy in infrared wavelengths	http://spitzer.caltech.edu
Suzaku	2005	JAXA/NASA	LELE (31°)	Study black holes and a wide variety of objects in X-ray wavelengths;	http://www.jaxa.jp/projects/index_e.html and then select Suzaku or ASTRO-EII
Akari	2006	JAXA	LELE (98.4°) Sun synchronous	All-sky survey in infrared wavelengths	http://www.jaxa.jp/projects/index_e.html and then select Akari
AGILE	2007	ISA	Low-Earth (2.5°)	Study pulsars, active galactic nuclei and other objects in X-ray and gamma-ray wavelengths	http://agile.rm.iasf.cnr.it/
Fermi gamma-ray	2008	NASA	Low-Earth (28.5°)	Study black holes, blazars, quasars, gamma-ray bursts in gamma-ray wavelengths	http://fermi.gsfc.nasa.gov/public/
WISE	2009	NASA	LELE (97.4°) (Sun-synchronous)	All-sky survey in mid-infrared wavelengths	http://wise.ssl.berkeley.edu/mission.html
Herschel space observatory	2009	ESA	Sun-Earth L_2 point	Study cold, star-forming areas and distant galaxies, survey the Universe in far-infrared wavelengths	http://www.herschel.caltech.edu
Astrosat	2011	ISRO	Low-Earth nearly equatorial	All-sky survey in ultraviolet, visible and hard x-ray wavelengths	http://meghnad.iucaa.ernet.in/~astrosat/home.html

GALEX Galaxy evolution explorer, *HST* Hubble Space Telescope, *WISE* Wide-field Infrared Survey Explorer, *SGR* Spektr-Rentgen-Gamma
ESA European Space Agency, *ISA* Italian Space Agency, *ISRO* Indian Space Research Organization, *JAXA* Japanese Space Agency, *NASA* National Aeronautics and Space Administration (USA)
HEHE high-Earth, high eccentricity, *HET* Heliocentric Earth-trailing, *LELE* Low-Earth, low eccentricity

Table 2.7. Past and current Sun-observing spacecraft

Name or acronym	Year of launch	Entity	Orbit type (inclination)	Purpose	Homepage
Ulysses	1990	ESA/NASA	Sun orbit, nearly polar	Characterize the heliosphere as a function of solar latitude	http://ulysses.jpl.nasa.gov/
SOHO	1995	ESA/NASA	Earth-Sun L_1 point	Characterize the Sun from its core to its corona; characterize the solar wind	http://sohowww.nascom.nasa.gov/
ACE	1997	NASA	Earth-Sun L_1 point	Measure composition of solar corona, solar wind and interplanetary particles	http://www.srl.caltech.edu/ACE/
TRACE	1998	NASA	Low Earth, Sun synchronous	Characterize the Sun's plasma, corona, photosphere and magnetic fields	http://sunland.gsfc.nasa.gov/smex/trace
Genesis	2001	NASA	Earth-Sun L_1 point	Characterize solar wind; capture some solar wind particles and return them to Earth	http://genesismission.jpl.nasa.gov/
RHESSI	2002	NASA	LELE (38°)	Characterize the basic physics of particle acceleration and explosive energy release in solar flares	http://hesperia.gsfc.nasa.gov/hessi/
Hinode	2006	JAXA and several others	LELE (polar orbit) Sun synchronous	Image the Sun in visible and X-ray wavelengths, measure magnetic fields and electrical currents near the Sun	http://solarb.msfc.nasa.gov/
STEREO	2006	NASA	Heliocentric; one spacecraft is in front of Earth and the other is behind Earth	Obtain 3-D images of coronal mass ejections; Characterize coronal mass ejections and their impact on Earth	http://stereo.jhuapl.edu
SDO	2010	NASA	Inclined geosynchronous	Characterize the Sun's influence on Earth; help us better understand the Sun-Earth connection	http://sdo.gsfc.nasa.gov/

ACE Advanced composition explorer, *RHESSI* Reuvenramaty high energy solar spectroscopic imager, *SOHO* Solar and heliospheric observatory, *STEREO* Solar terrestrial relations obervatory, *SDO* Solar dynamics observatory, *TRACE* Transition region and coronal explorer

ESA European Space Agency, *JAXA* Japanese Space Agency, *NASA* National Aeronautics and Space Administration (USA)

LELE Low Earth low eccentricity

Table 2.8. Recent Moon-observing spacecraft

Name or acronym	Year of launch	Entity	Orbit type (inclination)	Purpose	Homepage
SMART-1	2003	ESA	HMHE (polar orbit)	Test solar electronic propulsion; search for ice on the Moon	http://www.esa.int/export/SPECIALS/SMART-1/intdex.html
Kaguya	2007	JAXA	LMLE (polar orbit)	Obtain data on the lunar origin and evolution; develop technology for further lunar exploration	http://www.kaguya.jaxa.jp/index_e.htm
Chang'e-1	2007	CNSA	LMLE (near polar orbit)	Characterize lunar environment and surface regolith; test technology for further lunar exploration	http://nssdc.gsfc.nasa.gov/nmc/spacecraftDisplay.do?id=2007-051A
Chandrayaan-1	2008	ISRO	LMLE (polar orbit)	Produce a 3-D atlas of the Moon; conduct chemical and mineral- ogical mapping of the Moon	http://www.isro.org/chandrayaan/htmls/Home.htm
LRO	2009	NASA	LMLE (polar orbit)	Identify sites on the Moon close to potential resources for safe future robotic or human missions	http://lunar.gsfc.nasa.gov/
Chang'e-2	2010	CNSA	Moon orbit	Map the Moon; test soft landing technologies	http://nssdc.gsfc.nasa.gov/nmc/spacecraftDisplay.do?id=2010-050A
GRAIL	2011	NASA	LMLE (polar orbit)	Measure the Moon's gravitational field	http://moon.mit.edu/
Luna-Glob-1[a]	2012	RSA	Orbiter and lander	Collect data on the composition and temperature of the lunar surface; measure seismic activity on the Moon	http://www.russianspaceweb.com/luna_glob.html

GRAIL Gravity Recovery and Interior Laboratory, *LRO* Lunar reconnaissance orbiter, *SMART* Small missions for advanced research in technology

CNSA China National Space Agency, *ESA* European Space Agency, *ISRO* Indian Space Research Organization, *JAXA* Japanese Space Agency, *NASA* National Aeronautics and Space Administration (USA), *RSA* Russian Space Agency

HMHE high-Moon, high eccentricity, *LMLE* Low-Moon, low eccentricity

[a]See also Galimov (2005)

Table 2.9. Recent and successful spacecraft sent to Mars

Name or acronym	Year of launch	Study period (type)	Entity[b]		
MGS	1996	1997–2006 (orbiter)	NASA	Characterize Mars soil and topography; monitor daily and seasonal weather changes; characterize magnetic field	http://mars.jpl.nasa.gov/mgs/overview/
Mars Express	2003	2003-Current (orbiter)	ESA	Image surface at a resolution of 10 m/pixel; map minerals, determine global circulation	http://www.esa.int/SPECIALS/Mars_Express/index.html
MERM (Spirit and Opportunity)	2003	2004-Current (two rovers)	NASA	Search for and characterize a wide variety of surface rocks and soil which hold clues to past water activity	http://marsrovers.jpl.nasa.gov/home/index.html
MRO	2005	2006-Current (orbiter)	NASA	Characterize the existence and history of water on Mars	http://www.nasa.gov/mission_pages/MRO/main/index.html
Phoenix	2007	2008 (lander)	NASA	Study the history of water and habitability of potential life in the Martian Arctic ice-rich soil	http://phoenix.lpl.arizona.edu/mission.php
Curiosity[a]	2011	2012–2014 (rover)	NASA	Assess whether Mars ever was or is still today an environment able to support microbial life	http://marsprogram.jpl.nasa.gov/msl/

MERM Mars Exploration Rover Mission, *MGS* Mars Global Survey, *MRO* Mars reconnaissance orbiter, *ESA* European Space Agency, NASA National Aeronautics and Space Administration

[a]This is also called the Mars Science Laboratory or MSL

Table 2.10. Recent successful spacecraft that are observing other solar system bodies besides the Sun, Moon and Mars

Name or acronym	Year of launch	Study period	Entity	Purpose	Homepage
Cassini- Huygens	1997	2004-Current	NASA/ESA	Deliver Huygens lander to Titan; characterize the Saturn system	http://saturn.jpl.nasa.gov/index.cfm
Stardust	1999	2004	NASA	Characterize and collect dust of Comet 81P/Wild; collect interstellar dust	http://stardust.jpl.nasa.gov/home/index.html
Hayabusa	2003	2005–2007	JAXA	Characterize and collect a sample of minor planet 25143 Itokawa	http://www.jaxa.jp/projects/index_e.html and then select HAYABUSA or MUSES-C
Rosetta	2004	2008, 2010 and 2014	ESA	Characterize minor planets 2867 (Steins), 21 (Lutetia) and Comet 67P/Churyumov-Gerasimenko	http://www.esa.int/rosetta
MESSENGER	2004	2008-current	NASA	Characterize the surface, magnetic field and interior of Mercury	http://messenger.jhuapl.edu/the_mission/index.html
Deep impact	2005	2005	NASA	Characterize and release a projectile to Comet 9P/Tempel 1	http://deepimpact.umd.edu
EPOXI	2005	2010	NASA	Characterize Comet 103P/Hartley 2; observe transits of extra-solar planets their moons and rings	http://epoxi.umd.edu/
Venus express	2005	2006–2010[a]	ESA	Characterize Venus' atmosphere, surface, look for current volcanic activity	http://www.esa.int/SPECIALS/Venus_Express/index.html
Stardust-next	2006	2006-Current	NASA	Study Comet 9P/Tempel 1 and look for changes since 2005	http://stardustnext.jpl.nasa.gov
New horizons	2006	2015 (Pluto)	NASA	Characterize Pluto; characterize trans-Neptune objects	http://pluto.jhuapl.edu/
Dawn	2007	2011–2012 Vesta 2015 Ceres	NASA	Characterize Asteroid 4 (Vesta), and dwarf planet Ceres	http://dawn.jpl.nasa.gov/
Juno	2011	2016 and beyond	NASA	Improve our understanding of how Jupiter formed and evolved	http://science.nasa.gov/missions/juno/
NEAR	1996	2000–2001	NASA	Characterize Asteroid 433 (Eros)	http://near.jhuapl.edu

EPOXI contains letters from two acronyms namely *EPOCH* Extrasolar Planet observation and characterization and *DIXI* deep impact eXtended, investigation, *MESSENGER* Mercury surface space environment, geochemistry and ranging
ESA European Space Agency, *JAXA* Japanese Space Agency, *NASA* National Aeronautics and Space Administration (USA)
[a]from Wikipedia

Table 2.11. Recent spacecraft that will primarily collect data on phenomena beyond our Solar System

Name or acronym	Year of launch	Entity	Orbit (inclination)	Purpose	Homepage
WMAP	2001	NASA	Sun-Earth L_2 point	Study the universe as a whole; map the cosmic microwave background	http://map.gsfc.nasa.gov/
SWIFT	2004	NASA	LELE	Locate and study gamma ray bursts; scan half of the sky for new transient gamma-ray sources	http://swift.gsfc.nasa.gov/docs/swift/swiftsc.html
PAMELA	2006	Italy, Russia, Germary and Sweden	LEHE (70°)	Investigate cosmic radiation, dark matter, the origin and evolution of matter in our galaxy	http://pamela.roma2.infn.it/index.php
COROT	2006	ESA	LELE, polar orbit	Search for rocky planets outside our Solar System; search for acoustical waves moving across stars	http://www.esa.int/esaMI/COROT/index.html
THEMIS	2007	NASA	Several satellites, HEHE	Investigate auroras above Earth; collect data on the Sun-Earth connection	http://www.nasa.gov/mission_pages/themis/main/index.html
AAUSat II	2008	DSRT	LELE, Sun Synchronous	Study gamma ray bursts; test the student-built detector	http://www.space.aau.dk/aausatii/
IBEX	2008	NASA	HEHE	Characterize the interaction between the solar wind and the interstellar medium	http://www.ibex.swri.edu/index.shtml
Kepler	2009	NASA	Sun orbit, 1 year	Image a 105 square degree field near 19 h 20 m, 45°N and search for transiting extrasolar planets	http://www.nasa.gov/mission_pages/kepler/main/index.html
Planck	2009	ESA	Sun-Earth L_2 point	Image the anisotropies of the cosmic microwave background	http://www.rssd.esa.int/index.php?project=Planck
Ikaros	2010	JAXA	Sun orbit	Test a material that will be both a solar sail and a thin film solar cell	http://www.jspec.jaxa.jp/e/activity/ikaros.html
Gaia	2013	ESA	Lissajous-type around Sun-Earth L_2 point	Measure positions, distances and radial speeds of up to a billion stars	http://sci.esa.int/science-e/www/area/index.cfm?fareaid=26

COROT Convection rotation and planetary transits, *FASTSAT* Fast, affordable, science and technology satellite, *PAMELA* Payload for antimatter matter exploration and light-nuclei astrophysics, *WMAP* Wilkinson microwave anisotropy probe
DSRI Danish Space Research Institute, *ESA* European Space Agency. *JAXA* Japanese Space Agency, *NASA* National Aeronautics and Space Administration (USA)
HEHE high-Earth, high eccentricity, *HLE* Heliocentric, low-eccentricity, *LELE* Low-Earth, low eccentricity

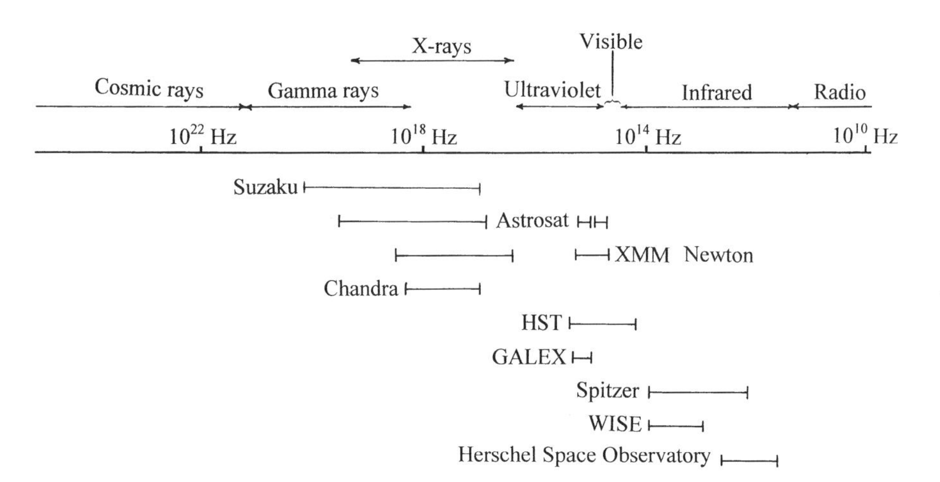

Fig. 2.4. Frequency ranges for several recent space telescopes. Data are from the websites listed in Table 2.6 (Credit: Richard Schmude, Jr.).

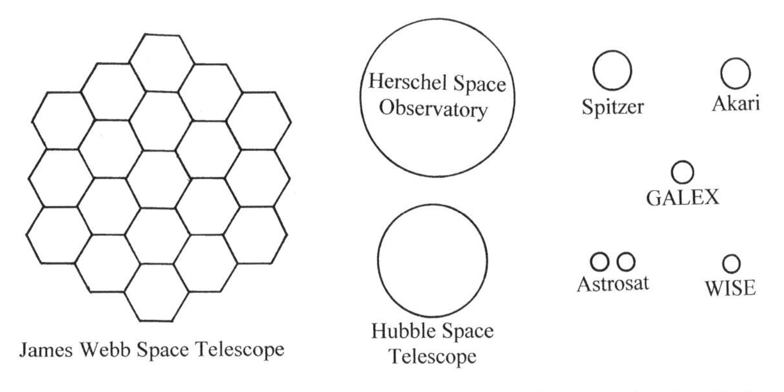

Fig. 2.5. Relative sizes of mirrors on a few recent and future space telescopes. The mirror diameters range from about 6.5 m for the James Webb Space Telescope down to 0.40 m for the Wide Field Infrared Survey Explorer (WISE). Data are from the websites in Table 2.6 (Credit: Richard Schmude, Jr.).

Table 2.12. Scientific objectives of the *Chang'e-1* lunar orbiter and relevant instruments. Data are from Huixian et al. (2005)

Objective	Instrument
Obtain a three dimensional (3-D) image of the lunar surface	Stereo camera, laser altimeter
Survey the thickness of the lunar regolith	Microwave radiometer
Determine the distribution of useful elements	Spectrometer imager, X-ray and gamma ray spectrometers
Explore the environment between the Moon and Earth	Particle detector, X-ray and gamma ray spectrometers

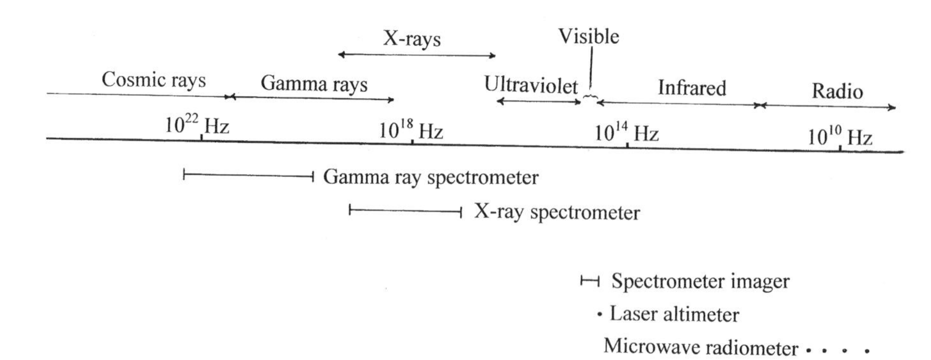

Fig. 2.6. Frequency ranges for different instruments on *Chang'e-1*. Data are from the sources listed in the text (Credit: Richard Schmude, Jr.).

Stereo Camera

The stereo camera resolved lunar features as small as 120 m across. Images were 60 km wide. It took three images at once, which were in the forward, nadir and backward positions. See Fig. 2.7. This camera allowed scientists to map the lunar surface and to gain information on the Moon's topography.

Chinese scientists constructed a map of the lunar surface from *Chang'e-1* images. It has a resolution of about 120 m.

Spectrometer Imager

This instrument covered visible and near-infrared wavelengths. It imaged a 25.6 km area. Its spatial resolution was 200 m. This is about the size of a city block. Scientists plan to use the data to look for different minerals on the Moon.

Laser Altimeter

The *Chang'e-1* orbiter had a laser altimeter. It sent pulses of near-infrared radiation with a wavelength of 1,064 nm. The pulses bounced off the lunar surface and reached a detector. This instrument collected topographic data. Chinese scientists obtained a topographical resolution of one meter from it.

Microwave Radiometer

The microwave radiometer served as both a microwave source and detector. It emitted microwaves (or short radio waves) with frequencies (in gigahertz – GHz) of 3, 7.8, 19.35 and 37. The respective wavelengths are 10, 3.8, 1.55 and 0.81 cm. The goal of this instrument was to collect data on the thickness and temperature of the lunar regolith.

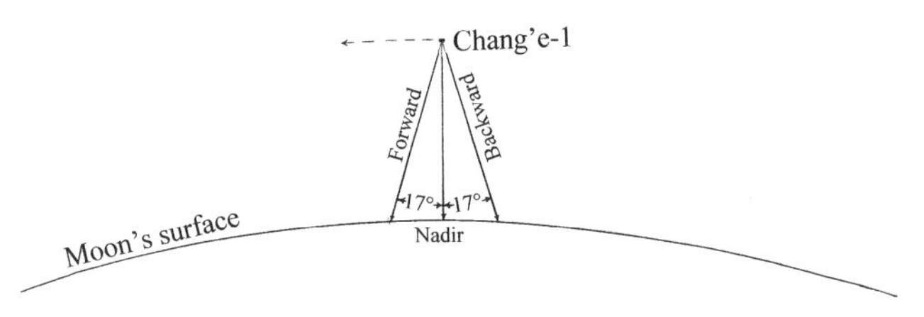

Fig. 2.7. The Stereo Camera on *Chang'e-1* recorded three images at once in the forward, nadir and backward directions. The center of the nadir image was 17° from the centers of the forward and backward images (Credit: Richard Schmude, Jr.).

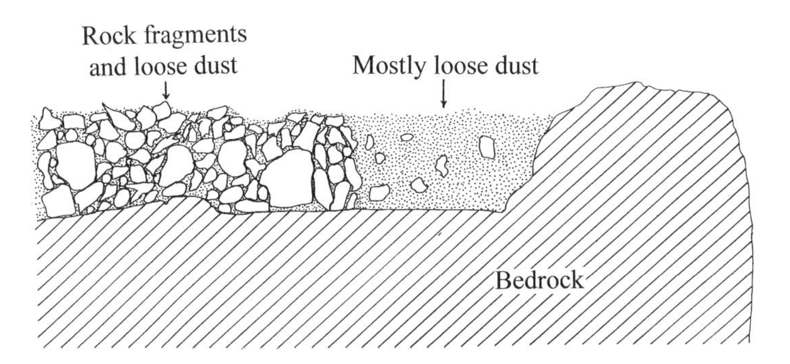

Fig. 2.8. The lunar surface contains either regolith or bedrock. The regolith is a mixture of rock fragments and loose dust. The regolith may be predominantly rock fragments (*far left*) or loose dust (*center*). As the surface cools down, the loose dust will cool off the fastest, and the bedrock will cool off the slowest (Credit: Richard Schmude, Jr.).

What cools off quicker after a hot day – a concrete sidewalk or dry beach sand? The beach sand cools off quicker. This is because small particles cool off faster than large ones. This also happens on the lunar surface. As rocks and other material rotate into daylight, they heat up, and as they rotate into night they cool off. Figure 2.8 shows a cross-section view of the Moon's surface. The top layer, called the regolith, is composed of dust and rock fragments. Most of the regolith fragments are not chemically bonded to one other. The underlying bedrock is composed of large areas of solid rock. Regolith, composed of loose dust, cools off quickly after sunset. Bedrock at the surface, however, cools off slowly. Therefore, the microwave radiometer data should enable scientists to determine where the exposed bedrock is and the depth of loose regolith.

Gamma and X-ray Spectrometer

The lunar surface emits gamma and X-ray radiation. Naturally, radioactive elements emit some of this radiation. Other elements, which are normally not radioactive, emit this radiation when struck with cosmic rays. See Fig. 2.9.

The purpose of the gamma ray and X-ray spectrometer is to yield information on the composition of the lunar surface and solar wind. It should also yield information on the distribution of uranium and other radioactive elements. How does it do this? Different elements emit different wavelengths of gamma and X-ray radiation. Emission occurs as a result of spontaneous emission by radioactive elements.

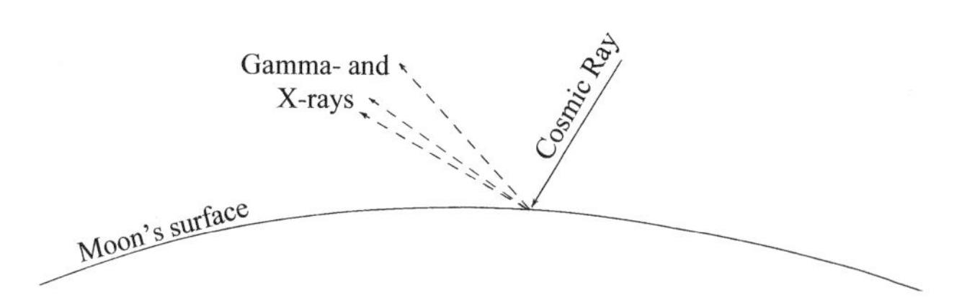

Fig. 2.9. When a cosmic ray strikes the Moon's surface, gamma- and X-rays are given off. The specific wavelength depends on the element being struck (Credit: Richard Schmude, Jr.).

It may also happen as a result of cosmic rays. Essentially, a cosmic ray hits the surface. An atom then emits specific wavelengths that are unique to the element. Scientists match the wavelengths given off with those given off by specific elements. In this way, they determine which elements are on the lunar surface.

Particle Detector

The particle detector measures the amount of protons and other subatomic particles with kinetic energies of between 4 and 400 MeV. It also detects other nuclei at similar energies. A 40 MeV is moving at about one third the speed of light. Therefore, the particle detector detects particles moving at very high speeds. Its purpose is to study the environment between Earth and the Moon. This information is needed to insure the safety of astronauts traveling to the Moon.

Chandrayaan – 1

India launched *Chandrayaan-1* on October 22, 2008, using a Polar Satellite Launch Vehicle (PSLV) rocket. It reached an elliptical transfer orbit like the one *Chang'e-1* followed. After repeated burns it left Earth's orbit and settled into a nearly circular lunar polar orbit. During the first 7 months of the mission, it traveled 100 km above the lunar surface. It moved at a speed of just over 2,900 km/hr and completed one orbit in 118 min. Every time *Chandrayaan-1* completed an orbit, the Moon rotated 1.08°. Therefore, the ground trace shifted 32.6 km at the equator. As a result this orbiter imaged nearly the entire lunar surface. On May 19 2009, Indian engineers raised the orbit to an altitude of 200 km. This enabled the instruments to study larger portions of the Moon. The Indian Deep Space Network (Indian DSN) facility near Bangalore was responsible for communication. Most of the information about the *Chandrayaan-1* orbiter is from Bhandari (2005), Krishna et al. (2005), Kumar and Chowdhury (2005a, b), Kamalakar et al. (2005), Goswami et al. (2005), Bhardwaj et al. (2005), Foing et al. (2005), from the *Chandrayaan-1* homepage, http://www.isro.org/chandrayaan/htmls/Home.htm and from the online version of *The Hindu* (India's national newspaper).

This satellite used solar arrays for power. They produced about 750 W of power. Much of this was stored in rechargeable lithium ion batteries. These were needed when the Moon blocked the Sun.

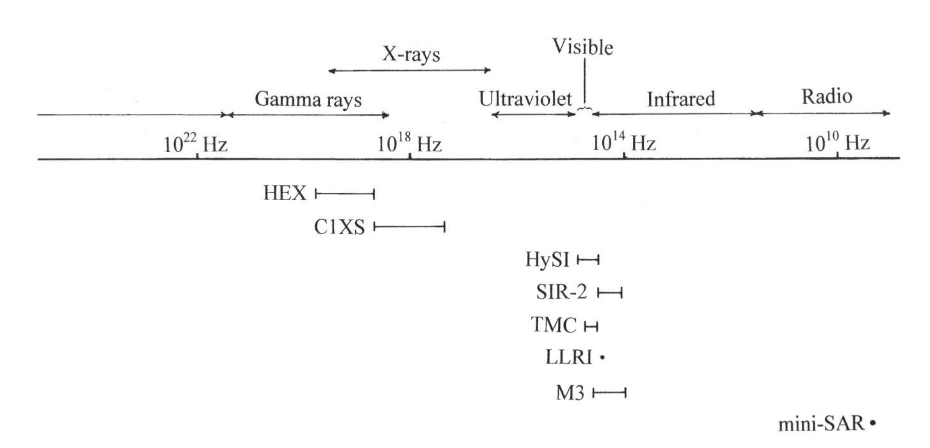

Fig. 2.10. Frequency ranges for different instruments on *Chandrayaan-1*. Data are from the sources cited in the text (Credit: Richard Schmude, Jr.).

Indian engineers built and designed most of the instruments on *Chandrayaan-1*. It carried out studies of the Moon, the solar wind and the solar wind's impact on the lunar environment. Figure 2.10 shows the frequency ranges which different instruments on *Chandrayaan-1* measured. Table 2.13 summarizes several instruments on board and their purpose. Let's examine some of these.

Terrain Mapping Camera (TMC)

The TMC imaged three regions of the Moon at once. Each was around 25° from the other. See Fig. 2.11. Scientists designed this camera to produce stereoscopic images. The camera's resolution was 10 arc-seconds. At an altitude of 100 km, this equals 5 m – about the size of a two-car garage. It also measured the heights of lunar features to an accuracy of 10 m – the height of a three-story apartment building.

One goal of the TMC team was to produce maps of the Moon's polar regions. The biggest problem was that very little light falls there. In essence, these areas are very dark. Astronomers reduced the problem of low-light levels by using the technique of pixel binning. Essentially, resolution was sacrificed to enhance sensitivity. With increased sensitivity they were able to image darker areas.

The TMC recorded about 70,000 images. One image set shows the 30 km crater De Roy in the forward (fore), nadir and backward (aft) positions. The shadows are different in these positions. Different viewing angles cause this. Indian astronomers constructed maps of most of the north and south polar regions of the Moon from TMC images. Over 80% of the areas south of 85°S and north of 85°N are in these maps. These are displayed on the website http://www.isro.org/Chandrayaan/htmls/Home.htm.

Table 2.13. Instruments on the *Chandrayaan-1* lunar orbiter along with their main purpose

Acronym	Instrument name (mass, power)	Purpose	Source
HEX	High energy X-ray spectrometer	Study the transport of volatile substances like radon	Sreekumar et al. (2009)
HySI	Hyper-spectral imaging Spectrometer (2.5 kg, 2.6 W)	Image the Moon's surface at different wavelengths	Kumar et al. (2009b)
SIR-2	Near infrared spectrometer (3 kg)	Collect infrared spectra of the Moon's surface	Mall et al. (2009)
TMC	Terrain mapping camera (6.3 kg, 1.8 W)	Map the lunar surface to a resolution of 5 ms	Kumar et al. (2009a)
LLRI	Laser altimeter (10 kg, 25 W)	Yield information on the Moon's topography	Kamalakar et al. (2009)
M^3	Moon mineral mapper spectrometer	Collect spectra of the Moon's surface that will yield information on the abundance and distribution of minerals	Pieters et al. (2009)
Mini SAR	Synthetic aperture radar (8.2 kg)	Search for evidence or water ice; characterize the lunar regolith	Spudis et al. (2009)
SARA	Neutral atom analyzer (4.4 kg, 17.1 W)	Image the surface and determine which atoms are sputtered; detect magnetic anomalies	Barabash et al. (2009)
Radom	Radiation dose monitor (0.098 kg, 0.1 W)	Monitor radiation in outer space and in lunar orbit	Dachev et al. (2009)
CIXS	X-ray spectrometer	Measure abundances of the major rock forming elements Mg, Al, Si, Ti, Ca and Fe	Grande et al. (2009)
MIP	Moon impact probe (34 kg)	Image the Moon's surface; analyze gas above the Moon's surface	Kumar and MIP Project Team (2009)

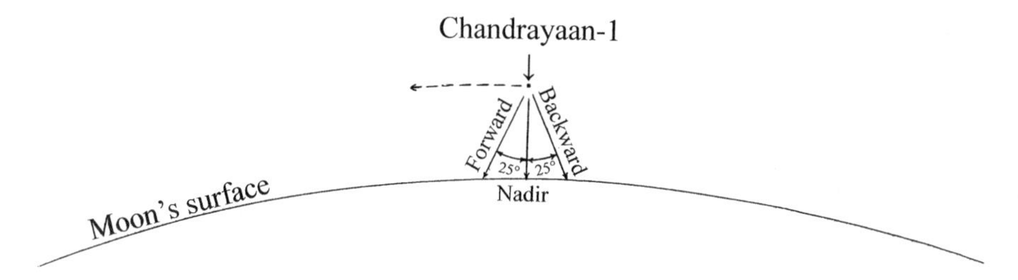

Fig. 2.11. The Terrain Mapping Camera on *Chandrayaan-1* recorded three images at once in the backward, nadir and forward directions. The center of the nadir image was 25° from the centers of the forward and backward images (Credit: Richard Schmude, Jr.).

Hyper-Spectral Imaging Spectrometer (HySI)

The HySI instrument imaged the lunar surface in 64 different wavelengths of light (around 415–920 nm). It used 64 different filters. With this instrument, astronomers collected both images and spectra.

Astronomers have released some results from the HySI. This instrument recorded 64 images of the 5-km crater Barrow H with each filter. The images show different colored materials on the inner crater wall. They also show tiny craters inside Barrow H. Some of these are only around 60 m across. This instrument also recorded images of the Moon's polar regions. Astronomers constructed color maps with these. Several images are displayed on the website http://www.isro.org/Chandrayaan/htmls/Home.htm.

Near Infrared Spectrometer (SIR-2)

The SIR-2 collected infrared spectra over the wavelength range 900–2,400 nm. It had a field-of-view of 0.127°. At an altitude of 100 km, this is equivalent to 220 m – about the size of a city block. The spectral resolution was 6 nm.

Data from this instrument should allow astronomers to distinguish between surfaces that contain mostly pyroxenes and those that contain mostly olivines. Pyroxene and olivine are two different minerals. Data from SIR-2 should yield information on the bright rays of Copernicus and Tycho.

Moon Mineral Mapper (M^3)

The M^3 instrument was a NASA supported guest instrument on *Chandrayaan-1*. Essentially, light entered a telescope, which focused it onto a spectroscope. A detector recorded the spectrum. This instrument covered wavelengths between 0.42 and 3.0 microns. Data from M^3 should enable astronomers to record spectra of different areas on the Moon.

In September 2009, NASA officials announced that the M^3 instrument detected trace quantities of water on the daylight side of the Moon (Sky and Telescope 2010a). See Figs. 2.12 and 2.13. According to one study, this water may come from high-energy protons, or hydrogen ions, which react with oxygen on the lunar surface. The trace amounts of water on the Moon may serve as an important resource for the occupants of a future Moon base.

Laser Altimeter (LLRI)

The purpose of the LLRI was to collect lunar topographical data. This was accomplished with a pulsed Nd-Yag laser. The wavelength was 1,064 nm. It sent short pulses every 0.1 s, which bounced off the Moon and back to the orbiter. Through careful measurement of the time that it took the light to leave and come back, scientists determined the distance to the ground. See Fig. 2.14. This instrument yielded altitudes to an accuracy of about 5 m.

This instrument also collected data on the Moon's interior. Essentially, the Moon's gravitational field is not uniform. It is stronger in some areas. Areas of stronger gravity are called mascons. A high concentration of iron or other dense substances may lead to a mascon. The LLRI yielded information on lunar mascons. By studying the small changes in lunar gravity, astronomers may learn more about the Moon's crust and interior.

According to the website http://www.isro.org/Chandrayaan, Indian astronomers have constructed an accurate map of Clavius Crater using LLRI data. It shows several hills on the crater floor. Astronomers have also constructed detailed topographical maps of the craters Bailly, Stebbins and Van't Hoff from LLRI data.

Fig. 2.12. Scientists assembled this mosaic of the Moon. They used images from the Moon Mineralogy Mapper on the *Chandrayaan-1* orbiter to make it. *Blue* areas indicate the presence of water. *Red* areas indicate the presence of the iron-bearing mineral pyroxene (Credit: ISRO/NASA/JPL-Caltech/Brown Univ./USGS).

High Energy X-ray Spectrometer (HEX)

The major goal of the HEX instrument is to search for evidence of volatile transport on the Moon. See Table 2.13. Here is how it works. Radon (^{222}Rn) is radioactive. It is normally a gas, but at very cold temperatures, it condenses into a solid. Therefore, when ^{222}Rn is released from the ground some of it may move along the lunar surface and end up freezing at the poles. If radon moves along the surface volatile transport would take place. The half-life of ^{222}Rn is 3.8 days, and hence, much of it decays quickly to other isotopes and ends up as the radioactive isotope lead – (^{210}Pb). If ^{222}Rn moves along the Moon's surface large amounts of ^{210}Pb should accumulate in the cold polar regions. If, on the other hand, ^{222}Rn does not move along the surface (no volatile transport) the Moon's polar region would not have a large amount of ^{210}Pb. See Fig. 2.15.

Sub-keV Atom Reflecting Analyzer (SARA)

The SARA instrument detected neutral atoms and molecules escaping from the Moon. It also detected ions. A neutral atom has a net electrical charge of zero, whereas an ion has a net positive or negative charge. Essentially, SARA was a neutral atom mass spectrometer.

How does SARA yield information on the Moon's surface? As it turns out nature helps. A stream of fast-moving charged particles bombards the Moon's surface. Some of these strike the surface and cause neutral atoms to be torn from their

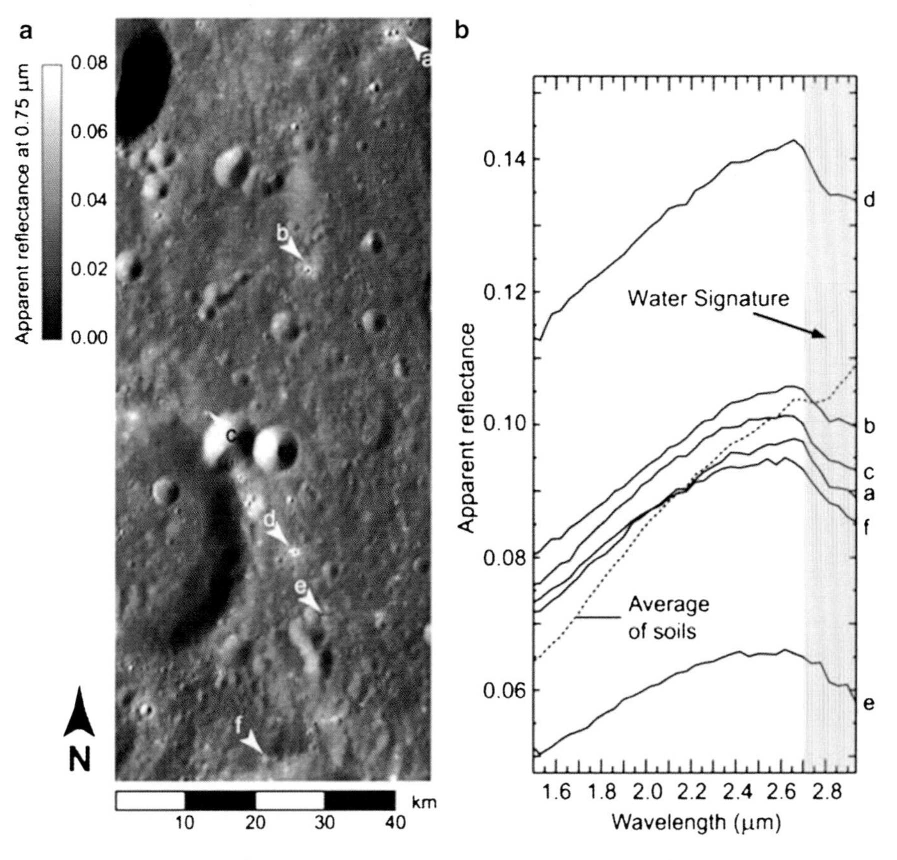

Fig. 2.13. The image on the *left* shows several lunar craters. The Moon Mineralogy Mapper on the *Chandrayaan-1* orbiter recorded spectra of the *white* regions with *arrows* (a–f). The spectra are shown on the *right*. The spectra are consistent with the presence of water in areas a-f (Credit: ISRO/NASA/JPL-Caltech/Brown Univ.).

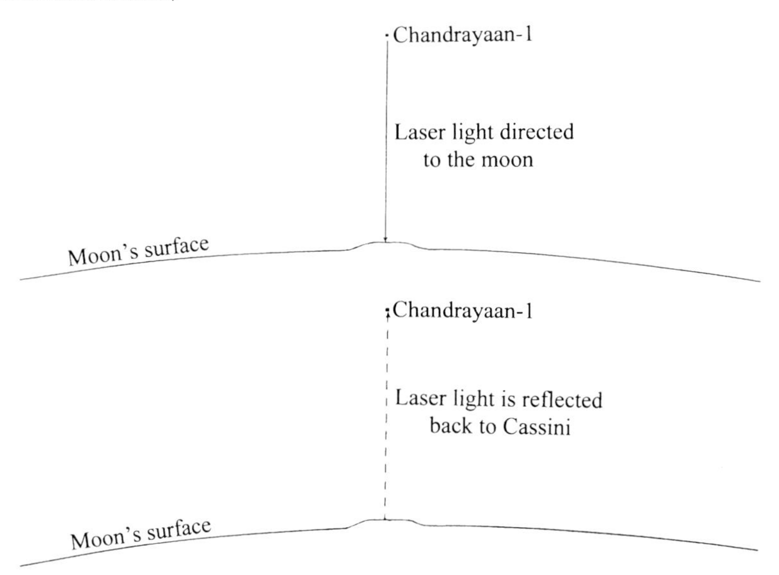

Fig. 2.14. The Laser Altimeter emits a pulse of radiation to the lunar surface. The pulse is then reflected back to *Chandrayaan-1*. By timing the pulse, astronomers are able to measure the distance between the orbiter and the lunar surface to an accuracy of 5 m (Credit: Richard Schmude, Jr.).

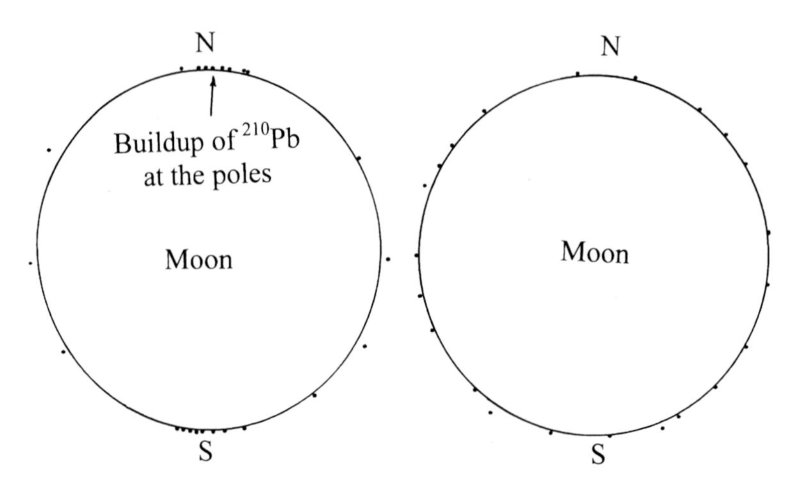

Fig. 2.15. Each dot represents ^{222}Rn (radon) and ^{210}Pb (lead) radioisotopes. If ^{222}Rn (radon) moves along the lunar surface, as in the *left* frame, it would accumulate near the poles because of the low temperatures. At the poles, the ^{222}Rn will turn into ^{210}Pb. If ^{222}Rn does not move along the surface, there would be no accumulation of ^{210}Pb at the poles, as is shown on the *right*. This is discussed further in Goswami et al. (2005) (Credit: Richard Schmude, Jr.).

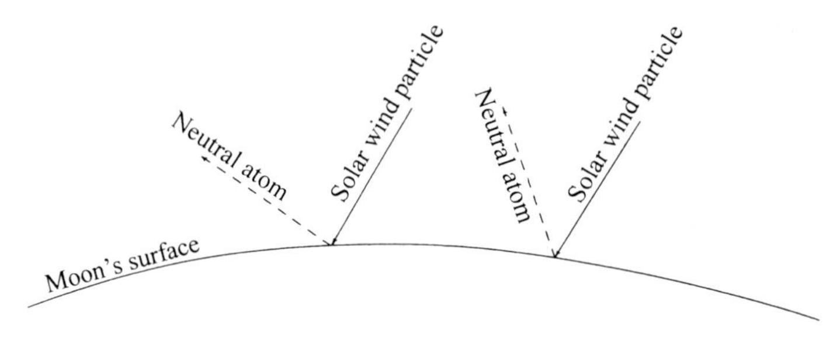

Fig. 2.16. A fast moving solar wind particle, like a proton, strikes the lunar surface. A neutral atom is released. A neutral atom is one that has the same number of protons and electrons. The Sub-keV Atom Reflecting Analyzer (SARA) on *Chandrayaan-1* detects neutral atoms. Their composition will give astronomers new insights concerning the elements on the lunar surface (Credit: Richard Schmude, Jr.).

crystal lattices and ejected. This is called sputtering. See Fig. 2.16. As a result of this, SARA was able to detect the escaping atoms. This allowed scientists to gain insights into the composition of the lunar surface.

A second objective of the SARA instrument was to characterize magnetic anomalies on the lunar surface. The Moon does not have a large magnetic field, like Earth, but it has weak fields that cover parts of its surface. These have strengths of up to about 0.05 Gauss. This is strong enough to prevent sputtering. Therefore, a drop in intensity of silicon atoms is evidence for a local magnetic field. See Fig. 2.17.

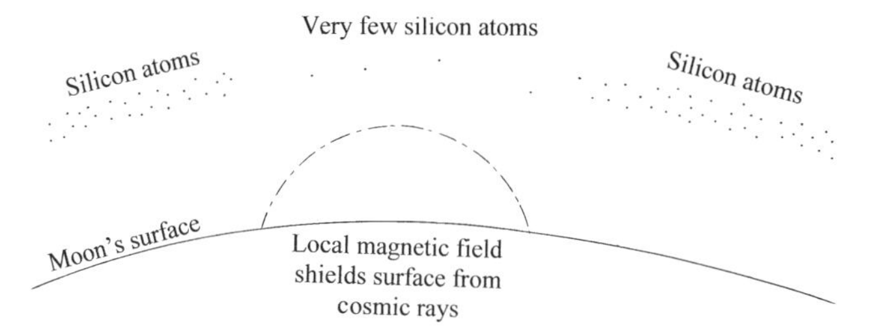

Fig. 2.17. Solar wind particles will generate a thin layer of silicon atoms far above the lunar surface. If a local magnetic field is present, however, most of the solar wind particles would not reach the surface. This will create a drop in the number of silicon atoms above the magnetic field. This information is based on Bhardwaj et al. (2005) (Credit: Richard Schmude, Jr.).

Chandrayaan-1 X-ray Spectrometer (C1XS)

The C1XS instrument package contained two instruments: an X-ray fluorescence spectrometer (LEX) and an X-ray solar monitor (XSM). The LEX instrument was designed to measure X-rays coming from the Moon's surface, and the XSM was designed to monitor X-rays coming from our Sun. The solar X-rays are needed for data calibration. The LEX instrument had a field-of-view of 14°. At an altitude of 100 km, this equals 25 km.

How did the C1XS measure element abundances? Essentially, it used X-ray fluorescence spectroscopy. Fluorescence is a process whereby radiation strikes a surface and less energetic radiation is emitted. In the case of the Moon, energetic radiation strikes the lunar surface. This causes electrons in the inner shells of atoms to shift positions. When this occurs X-rays are given off. Each element gives off specific wavelengths. For example, the elements silicon, aluminum, magnesium, calcium and iron give off X-rays with wavelengths (in nm) of 0.674, 0.795, 0.950, 0.307 and 0.174, respectively. (The wavelengths are computed from the energies listed in Lide (2008), p. 224). Therefore, if the C1XS detects lots of X-rays with a wavelength of 0.674 nm, silicon should be present.

Narendranath et al. (2011) used C1XS data to measure the soil composition in five spots on the Moon. They were near 29°S, 8.5°W. The average element abundances are: silicon (19%), aluminum (17%), calcium (8%), iron (7%) and magnesium (5%). Other elements make up the remaining 44%.

Miniature Synthetic Aperture Radar (miniSAR)

The miniSAR had an advantage over many of the other instruments on *Chandrayaan-1*. It did not require sunlight. Because of this, it collected data on the polar regions, which are in permanent darkness. Many astronomers believe that water ice is present in these areas because they do not heat up. This is how the miniSAR worked. It sent out radio waves with a frequency of 2.4 GHz. These interacted with the surface and bounced back to the orbiter. Many substances, including water ice, interacted with them in specific ways. Astronomers used miniSAR data to search for water ice.

Moon Impact Probe (MIP)

The MIP was a 34 kg package released soon after *Chandrayaan-1* began orbiting the Moon. As it fell, it spun and, as a result, was "spin stabilized." Small forces were unable to cause it to tumble. In essence, scientists knew the spin rate at all times. It collected data as it fell.

Summary of Chandrayaan-1

All eleven instrument packages functioned well during the mission. This satellite accomplished almost all of its objectives. It was the first satellite Indian engineers launched to the Moon. India joins a small group of countries that have placed its flag on the Moon. Unfortunately, the satellite started having difficulties in the spring of 2009 when one of its star sensors malfunctioned. A few months later the mission ended. Indian scientists are already working on a next generation lunar orbiter and lander – *Chandrayaan-2*. They are also analyzing data from *Chandrayaan-1* and have begun publishing their findings.

Hayabusa

Japanese engineers launched *Hayabusa* in May 2003, and it arrived at Asteroid 25143 Itokawa in September 2005. For the next 2 months it took images and recorded surface data. In November, the tiny lander, Minerva, separated from *Hayabusa* and made its slow descent to the surface. During this time, it collected microscopic dust grains. After several technical problems, scientists guided the return capsule back to Earth. It arrived in June 2010 in the Australian outback. The instruments on *Hayabusa* are summarized in Table 2.14. Figure 2.18 shows the frequency range that different instruments measured.

Hayabusa recorded images of all parts of Itokawa. These allowed scientists to measure its size, volume and surface area. Fujiwara et al. (2006) used the slight gravitational tugs of Itokawa to measure its mass. With the mass and volume they computed a density of 1920 ± 130 kg/m^3. This is about twice that of water but is only about 60% of the minerals making up the surface. Abe et al. (2006a) conclude that Itokawa has a porosity of 40%. This means that about 40% of this asteroid is empty space. Essentially, Itokawa is a rubble pile instead of a solid object. Over 500 boulders cover Itokawa's surface. Saito et al. (2006) measured their sizes. They found that the ratio of 20-, 10- and 5-m boulders is 1–8 to 500. A few characteristics of Itokawa are summarized in Table 2.15.

Hayabusa collected dust from Itokawa. This enabled scientists to carry out tests. Tsuchiyama et al. (2011) used a technique called microtomography to examine dust particles. They report grain sizes of up to 0.2 mm. Small grains are more abundant than larger ones over the 20–100 micron range. They report that the ratio of 80, 40 and 20 micron particles is 1 to 4 to 16. This group also reports a mean shape for several grains. The three dimensions are shown in Fig. 2.19. The shortest one, c, is 0.43 times the longest one, a. The intermediate dimension, b, is 0.71 times the longest one. This group also reports a mean density of 3430 kg/m^3 for dust

Table 2.14. Instruments on the *Hayabusa* spacecraft

Instrument (abbreviation)	Field of view	Source
Telescopic imaging camera (AMICA)	$5.8° \times 5.8°$	Saito et al. (2006)
Near infrared spectrometer (NIRS)	$0.1° \times 0.1°$	Abe et al. (2006b)
Laser ranging instrument (LIDAR)	5 m × 12 m	Abe et al. (2006a)
X-ray fluorescence spectrometer (XRS)	$3.5° \times 3.5°$	Okada et al. (2006)

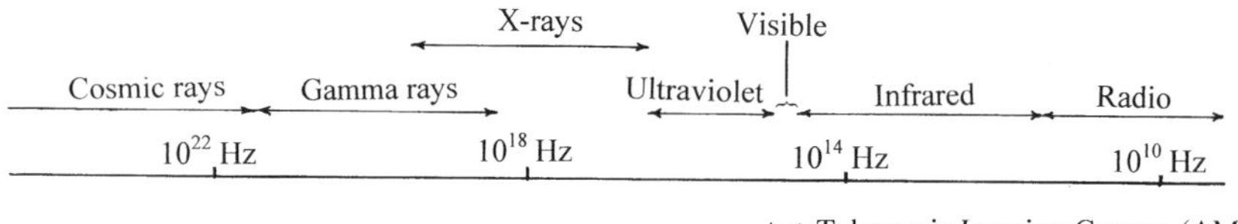

Fig. 2.18. Frequency range for different instruments on the *Hayabusa* spacecraft. Data are from Saito et al. (2006), Abe et al. (2006b), Abe et al. (2006a) and Okada et al. (2006) (Credit: Richard Schmude, Jr.).

Table 2.15. Characteristics of the Asteroid 25143 Itokawa. Most of these characteristics are based on data received from the *Hayakawa* spacecraft

Characteristic	Value	Source
Dimensions	535 m × 294 m × 209 m	Fujiwara et al. (2006)
Average density	1920 kg/m^3	Fujiwara et al. (2006)
Mass	3.51×10^{10} kg	Fujiwara et al. (2006)
Porosity	40%	Abe et al. (2006a)
Surface area	0.393 km^2	Demura et al. (2006)
Volume	0.0184 km^3	Demura et al. (2006)
Surface features	Areas covered with boulders, smoother areas, craters	Saito et al. (2006)
Rotation rate	12.1 h	Fujiwara et al. (2006)
Orbit	Inclination = 1.6°	Fujiwara et al. (2006)
	Eccentricity = 0.28	
	Perihelion distance = 0.95 au	
	Perihelion distance = 1.69 au	

grains. This is almost twice the bulk density of the asteroid. Table 2.16 lists characteristics of Itokawa's dust.

Nakamura et al. (2011) analyzed the mineral composition of 1,087 grains. They found that the majority of them (580) are olivine. Other minerals and grains are: feldspar (186), pyroxene (182), troilite (113), chromite (13), calcium phosphate (10) and iron-nickel (3).

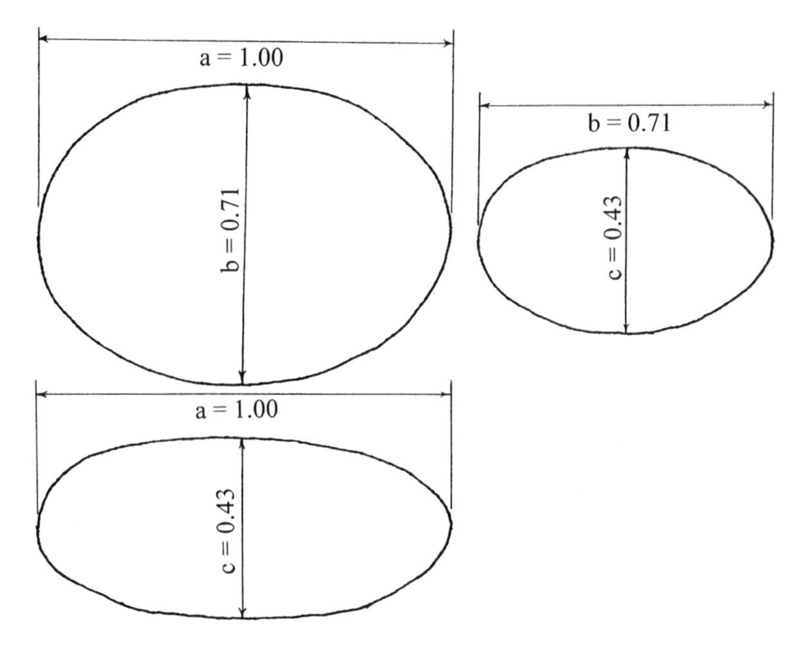

Fig. 2.19. The average shape of a dust grain from Asteroid 25143 Itokawa. This is based on data in Tsuchiyama et al. (2011) (Credit: Richard Schmude, Jr.).

Table 2.16. Characteristics of the dust collected near Itokawa

Characteristic	Value	Source
Size	up to 0.2 mm	Nakamura et al. (2011)
Shape	a = 1.00, b = 0.71, c = 0.43	Tsuchiyama et al. (2011)
Density	3430 kg/m^3	Tsuchiyama et al. (2011)
Composition (minerals)	Olivine, feldspar, pyroxene, troilite, chromite, calcium-phosphate, iron nickel	Tsuchiyama et al. (2011)
Average porosity	1.4%	Tsuchiyama et al. (2011)

Hubble Space Telescope

Have you ever looked through a large telescope? If so, you probably noticed that the image was a little blurry. Turbulence and air currents in our atmosphere cause this. Furthermore, our atmosphere acts as a weak prism. Essentially, it bends the colors of light in different directions. These factors degrade the image. It is for this reason that astronomers began thinking about a space telescope before the space age.

NASA and other organizations have launched several space telescopes covering much of the electromagnetic spectrum. See Table 2.6. The most well known of these is the Hubble Space Telescope (HST). One reason for this is that it operates at visible wavelengths. This is the same light we see. According to the website (http://spacetelescope.org/) as of July 2011 Hubble has made over 1 million observations. The data that HST has collected has helped scientists publish 9,400 scientific

Table 2.17. Characteristics of the Hubble space telescope (HST). Much of the information is from http://spacetelescope.org

Characteristic	Value
Length	13.2 m
Diameter	4.2 m
Mass at launch	11,110 kg
Orbit	569 km, circular
Orbital inclination	28.5°
Launch date	April 24, 1990
Launch vehicle	space shuttle discovery
Date placed in orbit	April 25, 1990
Wavelengths capability of the HST	Ultraviolet, visible and infrared
Number of observations made as of mid 2011	Over 1,000,000
Number of images made as of late 2009	570,000
Amount of data produced as of late 2009	39 TB

Table 2.18. A summary of HST servicing missions. Data are from Blades (2008) and the HST website

Servicing mission (acronym)	Date	Shuttle used
Service Mission 1 (SM1)	December 1993	Endeavour
Service Mission 2 (SM2)	February 1997	Discovery
Service Mission 3A (SM3A)	December 1999	Discovery
Servicing Mission 3B (SM3B)	March 2002	Columbia
Servicing Mission 4 (SM4)	May 2009	Atlantis

papers. The telescope was named after the American astronomer Edwin Powell Hubble (1889–1953) who discovered that the universe is expanding. Table 2.17 lists a few characteristics of this telescope.

The HST in 2011 is a little like the "Six Million Dollar Man" – a popular TV character in the 1970s. Many of its original parts have been replaced with new ones. The four HST servicing missions are summarized in Table 2.18. The purpose of these was to replace or repair broken equipment and to place the HST in a higher orbit. The orbit change was needed because the telescope experiences small drag forces that cause it to fall closer to Earth. Without the boost to a higher orbit, it would spiral back to Earth. Space shuttle astronauts carried out all of the servicing missions. The Canadian-built robotic arm was also of great value in these missions. Much of the material in this section is from the website http://spacetelescope.org/, and Petersen and Brandt (1998).

The HST contains several features that enables it to function properly and to be repaired easily. It contains a support structure that holds equipment in place. Instruments in this structure are arranged in a modular format. This insures that they may be removed easily. The HST also has a device which blocks out stray light. Insulation protects the equipment form the extreme temperature changes in outer space. This telescope moves into Earth's shadow and as a result, the temperature

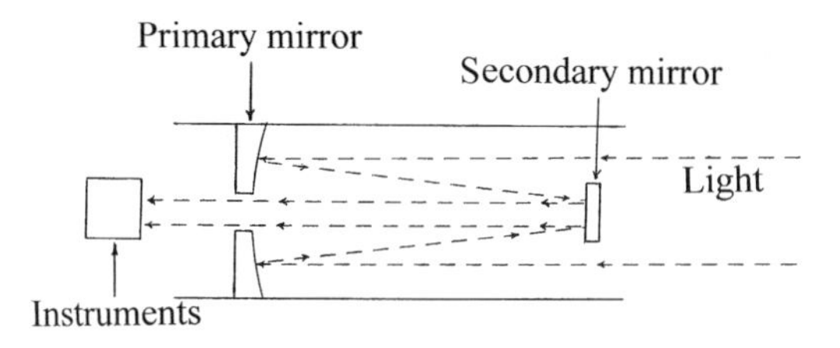

Fig. 2.20. Light enters the Hubble Space Telescope and then bounces off the primary mirror. This light then strikes the secondary mirror and moves back through the hole in the primary mirror to the instruments. This information is from the HST website (Credit: Richard Schmude, Jr.).

changes by about 100°C. In addition to these items, six equipment systems on Hubble are essential to its operation. These deal with optics, power generation/storage, maneuverability and pointing, computer/data storage, communications and science. This equipment will be described here.

Optical System

The HST contains a 2.4 m primary mirror and a 0.34 m secondary mirror. The primary mirror has a focal length of 57.6 m and, hence, its focal ratio is f/24. It is coated with aluminum, which is the same element that is on the mirrors of most Newtonians. On top of the aluminum is a protective layer of magnesium fluoride. The optical system is of the Ritchey-Crétien type. Essentially, light comes into the telescope and reflects off the primary mirror, which then moves to the secondary mirror. Light reflects off that and moves through the central hole in the primary mirror to the instruments. See Fig. 2.20.

The mirror has one problem. It does not focus all of the light at the same point. This is called chromatic aberration. This is the same problem that many inexpensive refractor telescopes have. The problem came about because the mirror was manufactured to the wrong specifications. Since engineers know the exact shape of the mirror, they are able to make modifications that correct the problem. Figures 2.21 and 2.22 show images of the planets Mars and Jupiter taken with the HST. The images show clearly that engineers have solved the problem.

Power Generation/Storage System

The HST has two solar arrays with a total area of 37 m^2. They supply over 5 kW of power – enough to power a small home. The solar arrays charge the six nickel-hydrogen rechargeable batteries. These serve as a source of power. When the HST moves into Earth's shadow, the arrays are not able to generate power. It is during this time that stored energy becomes essential.

Maneuverability and Pointing System

As of late 2009, the HST had imaged over 29,000 targets without anyone pushing it. Astronomers were able to steer this instrument with reaction wheels. We will

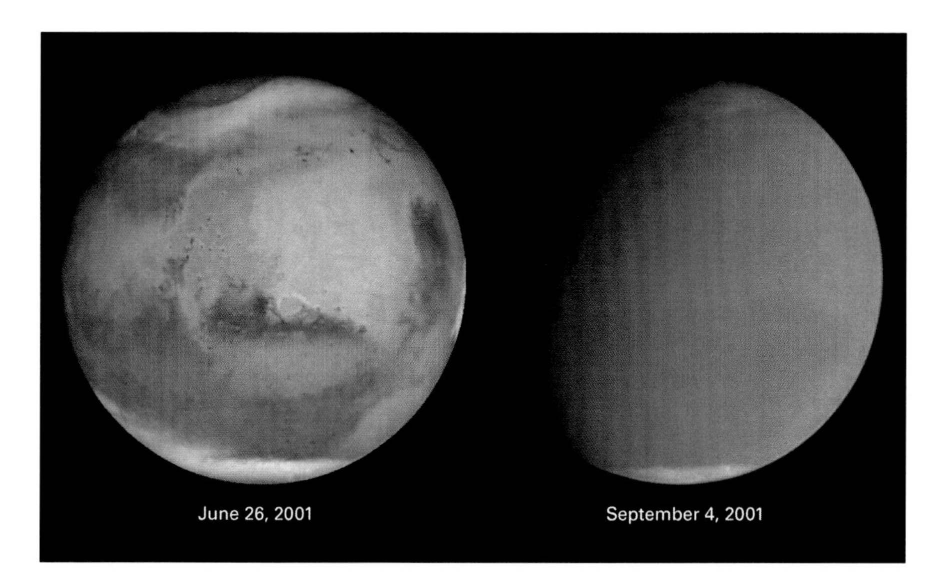

Fig. 2.21. Two images of Mars taken with the corrected optics of the Hubble Space Telescope. The June 26, 2001, image was taken before a major dust storm. Note the sharp detail and high resolution of the HST image. The other image shows Mars shrouded in a major dust storm (Credit: NASA, James Bell, Michael Wolf and the Hubble Heritage Team).

Fig. 2.22. An image of Jupiter taken with the corrected optics of the Hubble Space Telescope (Credit: NASA/ESA).

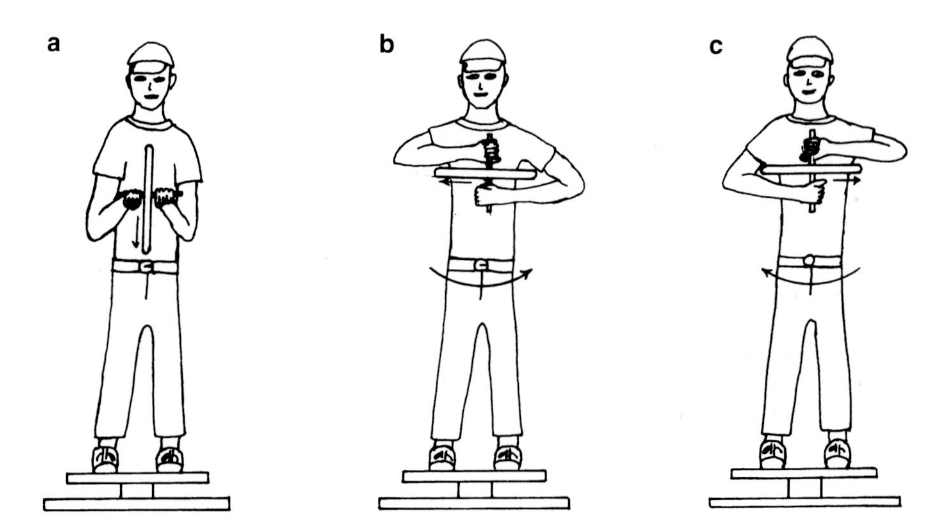

Fig. 2.23. A boy holds a spinning bicycle wheel and stands on a lazy Susan. Frame a: the boy does not spin because he holds the bicycle wheel in such a way that it does not affect his angular momentum around the rotational axis of the lazy Susan. Frame b: the boy spins in the opposite direction in which the bicycle wheel spins because of the conservation of angular momentum. Frame c: once again, the boy spins in the opposite direction in which the bicycle wheel spins. Therefore, the boy controls the direction he spins by how he holds the wheel (Credit: Richard Schmude, Jr.).

explain how this is done. Figure 2.23 shows a boy standing on a lazy Susan, a device that can spin. Its rotational axis is perpendicular to the ground. The boy is holding a spinning bicycle wheel so that its rotational axis is perpendicular to that of the lazy Susan. In this example, the boy and bicycle wheel are analogous to the HST and a reaction wheel, respectively.

The boy does not spin in Fig. 2.23a because the rotational axis of the spinning wheel is perpendicular to that of the lazy Susan. Essentially, he does not spin because of how he is holding the wheel. When he turns the wheel as is shown in Fig. 2.23b, he spins. He spins because of what is called the conservation of angular momentum. Essentially, angular momentum is the product of the mass and angular speed of a rotating object. In Fig. 2.23a, the total angular momentum around the axis of the lazy-Susan is zero. The total angular momentum in this case is the sum of the boy's angular momentum and that from the bicycle wheel. This is also the case in Fig. 2.23b. In Fig. 2.23b, the wheel has a non-zero angular momentum with respect to the axis of the Lazy Susan and, hence, the boy has an equal but opposite angular momentum to what he is holding. Therefore, the boy spins in the opposite direction of the wheel. As a result, the sum of the angular momentum of the boy and the wheel equals zero.

Now, when the boy flips the wheel as is shown in Fig. 2.23c, he spins in the opposite direction. The boy may also hold the wheel at a 45° angle and he will spin but at a slower rate than in Fig. 2.23b or c. Therefore, the boy controls how fast he spins by how he holds the wheel. In a similar way, astronomers control HST with the reaction wheels.

The HST must find two guide stars near its target before data collection commences. Once suitable stars are found, the Fine Guidance Sensors (FGSs) lock onto them. One of the tasks of the FGSs is to keep the HST pointed at its target.

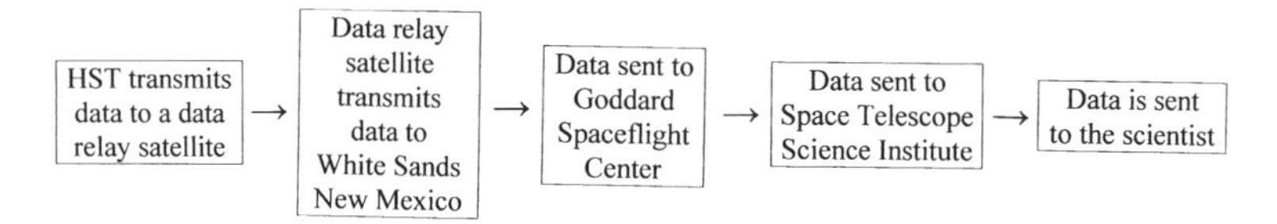

Fig. 2.24. The sequence of data transmission for the Hubble Space Telescope (Credit: Richard Schmude, Jr.).

The FGSs keep it pointed at the target to within 0.01 arc-seconds. This is one reason why HST images are sharp. A FGS unit is the size of a washing machine.

Three rate sensor units (RSUs) are also in the HST. Each of these has two gyroscopes. These spin at a rate of 19,200 rpm. The RSUs are needed to point the HST. NASA replaced all of the gyroscopes in the fourth servicing mission (SM4).

Computer/Data Storage System

The HST has a computer with ample memory to store data. Essentially, one instrument collects data that is stored on the computer. Later, it is sent back to Earth. This computer also receives detailed instructions that cover all operations of the HST including pointing, calibration activities, the sequence of observations to be carried out, and instructions for keeping the mirror pointed away from very bright objects.

Communications System

The HST has two radio dishes that receive and transmit data. All data is transmitted through radio signals. Data is generally transmitted from HST to one of several of NASA's *Tracking and Data Relay Satellites.* These are in geostationary orbits. They relay the data to White Sands, New Mexico. Afterwards, the data is sent to Goddard Spaceflight Center in Greenbelt, Maryland, the Space Telescope Science Institute at Johns Hopkins University in Baltimore, Maryland, and finally to the appropriate observer. The sequence of data transmission is shown in Fig. 2.24.

Scientific Instrument System

The HST has several scientific instruments. Five of these are the Advanced Camera for Surveys (ACS), Cosmic Origins Spectrograph (COS), Near Infrared Camera and Multi-Object Spectrometer (NICMOS), Space Telescope Imaging Spectroscope (STIS) and the Wide Field Camera 3 (WFC 3). In addition to these, the Fine Guidance Sensors (FGS) are often used as an instrument. Figure 2.25 shows the frequency range for HST's instruments. We will describe each of these, but before this, we will describe what a spectrum is and how astronomers use it.

A spectrum shows a pattern of light that is the result of a substance interacting with light. Two types of spectra are called absorption and emission spectra. They are often presented as graphs of light intensity versus wavelength. Figure 2.26

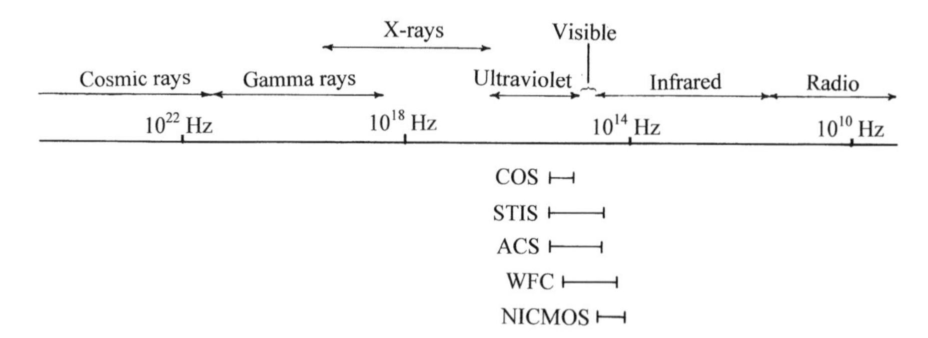

Fig. 2.25. Frequency ranges for different instruments on the Hubble Space Telescope (HST). Data is from the HST home page (Credit: Richard Schmude, Jr.).

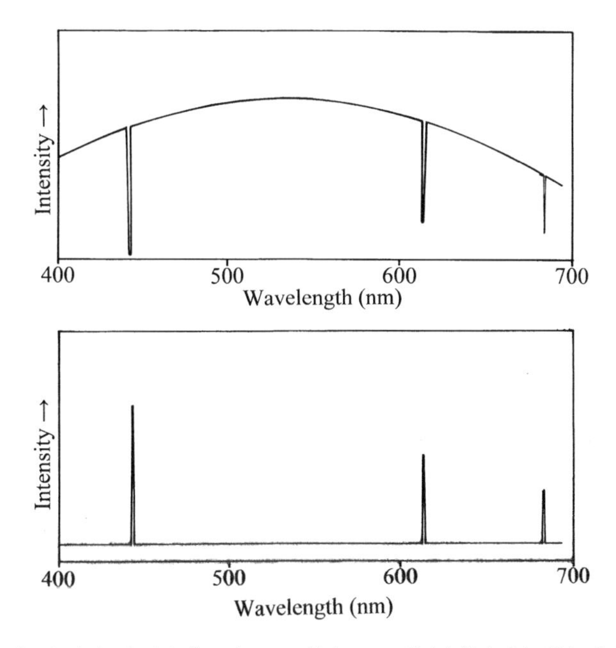

Fig. 2.26. An absorption (*top*) and emission (*bottom*) spectrum of hydrogen in visible light (Credit: Richard Schmude, Jr.).

shows an emission and an absorption spectrum of hydrogen. In the emission spectrum the light intensity is near zero except for three specific wavelengths. The visible light hydrogen emits corresponds to the three emission peaks. The hydrogen absorption spectrum has intensity minima at the wavelengths in Fig. 2.26. Absorption causes the minima. Under certain conditions, hydrogen emits light, and under others, it absorbs light.

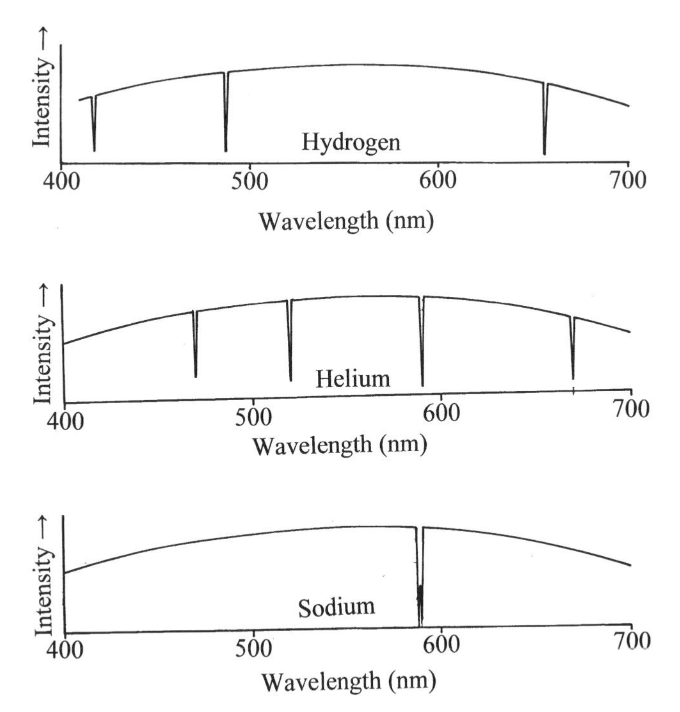

Fig. 2.27. Absorption spectra of hydrogen, helium and sodium. Note that each of these elements absorbs different wavelengths of light (Credit: Richard Schmude, Jr.).

Each element, ion and compound produces its own unique spectrum. See Fig. 2.27. Because of this, one may identify an element in a distant object by looking at its spectrum. For example, when the pattern of emission (or absorption) lines matches up with that of hydrogen, it means that this element is present. See Fig. 2.28. Astronomers can use spectra to identify elements, ions and compounds. Table 2.19 summarizes some of the other information that astronomers may obtain from spectra.

Advanced Camera for Surveys (ACS)

The ACS images an object and records its spectrum. Essentially, it is two instruments in one. This camera breaks up the light from a target into individual wavelengths. It can also block out unwanted light. This is especially useful for studying planetary satellites or a faint object near a bright star. The ACS also contains a device that blocks out unwanted visible light. In this way, astronomers are able to focus just on the ultraviolet light. For example, it is an ideal instrument to study Jupiter's aurora since it cuts out the visible light this planet reflects.

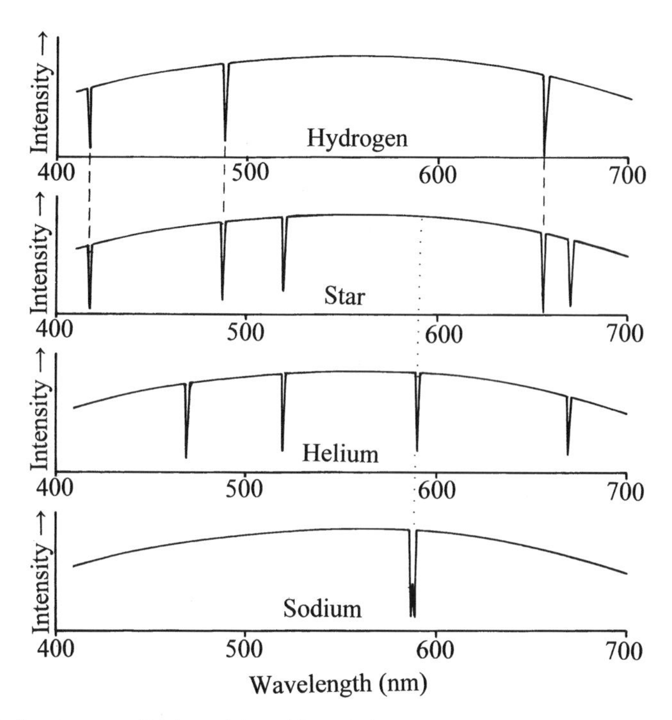

Fig. 2.28. Absorption spectrum of the elements hydrogen, helium and sodium along with that of a star. Note that the three absorption features of hydrogen match up with those in the spectrum of the star (*dashed line*). This shows the star has hydrogen. The absorption features in the spectra of helium and sodium do not match up with the star (*dotted lines*). Therefore, one may conclude that those elements are not in the star's spectrum (Credit: Richard Schmude, Jr.).

Table 2.19. Information that may be obtained from spectra

Information	What we should do
Identify an element, ion or compound	Match up the target's spectrum with that of an element, ion or compound. A match means the element, ion or compound is present
Temperature measurement	Measure the intensity of light given off at different wavelengths; match up the intensities with a blackbody at different temperatures. The one that matches the spectra best is the object's temperature
Rotation rate of a star	Measure the width of an emission line. The wider the line, the faster the rotation
Measure magnetic field strength	Measure small shifts in the positions of absorption (or emission) lines caused by nearby magnetic fields. These shifts are the result of the Zeeman Effect. The magnitude of the shifts is used to measure the strength of the magnetic field
Isotopic abundance	Measure the rotational transitions in high-resolution spectra. Isotopes will have different rotational spectra. The intensity of rotational transitions is used to determine isotopic abundances
Radial velocity	Measure how far the absorption (or emission) lines shift. Use the Doppler shift equation to determine the radial velocity
Extrasolar planet or other unseen companion	Record spectra over a period of months or years. Look for periodic shifts in the wavelength of absorption (or emission) lines. These shifts will reveal the period of revolution and the mass of the stellar companion in terms of the star

Cosmic Origins Spectrograph (COS)

The COS collects spectra. It has two channels that are sensitive to different wavelengths of ultraviolet light. One is sensitive to wavelengths between 115 and 177 nm, and the other is sensitive to those between 175 and 300 nm. This instrument is about the size of a hot water heater.

Near Infrared Camera and Multi-Object Spectrometer (NICMOS)

The NICMOS records both images and spectra. It has a field-of-view of 52×52 arc seconds – slightly smaller than the Ring Nebula. It detects objects as faint as 25th magnitude. This is about 10,000 times fainter than Pluto. Unlike the other HST instruments, the NICMOS detector is cooled to 80°K. This is needed to reduce noise caused by heat. A large block of nitrogen ice serves as the coolant. Like the ACS, this instrument is able to block out unwanted light.

Space Telescope Imaging Spectrograph (STIS)

As the name implies, STIS takes an image and records spectra. It may also record spectra of different parts of an extended object. For example, this instrument can record the spectra of the nucleus and the spiral arms of a galaxy. Like the ACS and NICMOS, STIS can block out unwanted light. This instrument has a field-of-view of 50×50 arc seconds. It is able to detect objects as faint as magnitude 28.5. This is about the brightness of a 100-W light bulb at a distance of 8 million km. It has several different gratings. This allows the astronomer to choose the resolution and the wavelengths of light to be included in the spectrum.

Wide Field Camera 3 (WFC 3)

The WFC 3 is about the size of an upright piano. It records an image at a specific wavelength. This instrument has two channels that are sensitive to 200–1,000 nm and to 850–1700 nm. The first channel has a field-of-view of 160 arc seconds by 160 arc seconds, which is about twice the size of the Ring Nebula (M57). The detector has an array of 4,096 by 4,096 pixels. Therefore, each pixel has a resolution of 0.04 arc seconds. This is one reason why HST images are sharp. The second channel has a field-of-view that is about half that of the first one. It also has a lower resolution. Like the other scientific instruments on the HST, the WFC 3 has optical components that correct for the spherical aberration of the primary mirror.

Fine Guidance Sensors (FGSs)

As mentioned earlier, the primary function of the FGSs is to keep the HST pointed at its target. A second function is to yield information on the positions of stars (astrometry). The HST Fine Guidance Sensors are able to yield more accurate star positions than those obtained from the ground because the atmosphere does not smear out the images.

Cassini

NASA launched the *Cassini* Saturn orbiter on October 15, 1997 with a Titan Centaur rocket from Cape Canaveral, Florida. At launch, this spacecraft had a mass of 5650 kg – twice that of a large sport utility vehicle. It is named after the Italian astronomer Jean Dominique Cassini, who discovered four of Saturn's moons (Iapetus, Rhea, Dione and Tethys). It took this spacecraft almost 7 years to reach its destination. *Cassini* fell into an orbit around Saturn on June 30, 2004. Along the way, it received gravity assists from Venus, Earth and Jupiter. This spacecraft also passed about 1.6 million km from asteroid 2685 (Masursky) on January 23, 2000. It also imaged Jupiter. See Fig. 2.29.

The main objective of *Cassini* is to collect data on Saturn, its rings and satellites. Because of this, it continues to orbit that planet and make flybys of the moons. The necessary orbital changes are made with an onboard rocket. Starting in 2008, Cassini started its extended mission. It may have enough power to last until 2015 or beyond.

Three Radioisotope Thermoelectric Generators (RTGs) supply power. These are needed to operate the equipment and to maintain the proper temperature for each instrument. Chapter 1 describes RTGs.

Fig. 2.29. The narrow-angle camera onboard *Cassini* took several close-up images of Jupiter. Astronomers then constructed a mosaic of Jupiter from these images, which is shown here (Credit: NASA/JPL/Space Science Institute).

Table 2.20. Instrument packages on *Cassini* along with some technical data

Acronym	Instrument name	Mass (kg)	Power (Watts)	Source
ISS	Imaging science subsystem	56.9	45.6	Porco et al. (2005b)
VIMS	Visual and infrared mapping spectrometer	–	–	Brown et al. (2005)
CIRS	Composite infrared spectrometer	–	–	Flasar et al. (2005b)
UVIS	Ultraviolet imaging spectrograph	15.6	8	Esposito et al. (2005a)
CAPS	Cassini plasma spectrometer	23.0	16.4	Young et al. (2004)
INMS	Ion and neutral mass spectrometer	10.3	23.3	Waite et al. (2004)
MIMI	Magnetospheric imaging instrument	28.1	20.3	Krimigis et al. (2004)
MAG	Magnetic field investigation	8.8	12.6	Dougherty et al. (2004), Narvaez (2004)
RPWS	Radio and plasma wave science	37.7	16.4	Gurnett et al. (2004)
CDA	Cosmic dust analyzer	17	12	Srama et al. (2004)
–	Cassini radio science	–	–	Kliore et al. (2005)
RADAR	RADAR: Cassini Titan Radar mapper	43.3	195	Elachi et al. (2005)

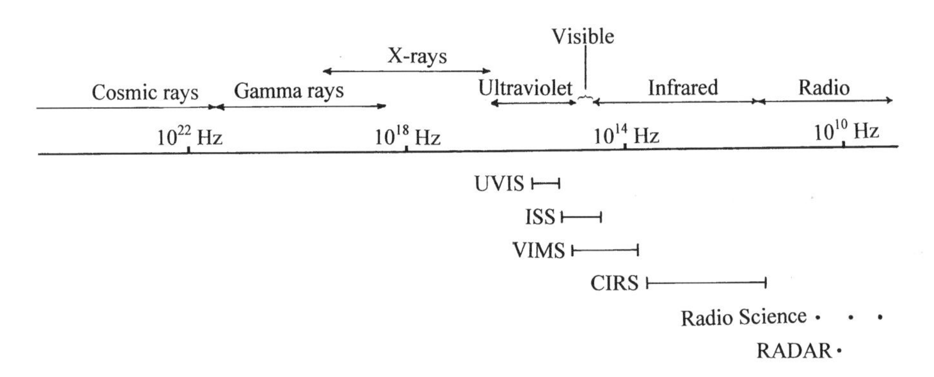

Fig. 2.30. Frequency ranges for different instruments onboard *Cassini.* Data are from the *Cassini* home page and from a series of papers published in the journal *Space Science Reviews,* (2005) volume 115, pp. 1–518 (Credit: Richard Schmude, Jr.).

Cassini has twelve scientific instrument packages onboard along with the Huygens Titan Lander. Table 2.20 lists instruments and technical information. Figure 2.30 shows which frequencies they detect. Technical characteristics of several instruments are summarized in Tables 2.21 and 2.22. A few of these will be described.

Table 2.21. Characteristics of the ultraviolet, visible and infrared instruments on *Cassini*; in this table, FOV = field of view in degrees, SR = spatial resolution in arc seconds/pixel

Acronym	Subsystem, mode or channel	FOV	SR	Purpose
ISS[a]	Wide angle camera (WAC)	3.5	12	Record images at wavelengths of between 0.38 and 1.1 mm
"	Narrow angle camera (NAC)	0.35	1.2	Record images at wavelengths of between 0.2 and 1.1 mm
VIMS[b]	Visual channel (VIMS-VIS)	1.8[c]	34	Collect images and spectra at wave-lengths of between 0.3 and 1.05 mm
"	Infrared channel (VIMS-IR)	1.8	52	Collect images and spectra at wave-lengths of between 0.85 and 5.1 mm
UVIS[d]	Far ultraviolet	–	–	Collect spectra at wavelengths of between 56 and 118 nm
"	Near ultraviolet	–	–	Collect spectra at wavelengths of between 110 and 190 nm
"	Hydrogen-deuterium absorption cell	–	–	Measure the deuterium-to-hydrogen ratio in the atmospheres of Saturn and Titan
CIRS[e]	Polarizing interferometer	0.22	–	Collect infrared spectra at wavelengths of between 16.7 and 1000 mm
"	Mid-IR Michelson Interferometer	0.016	–	Collect infrared spectra at wavelengths of between 7.1 and 16.7 mm

[a]Porco et al. (2005b)
[b]Brown et al. (2005)
[c]This is the normal FOV; this instrument had a maximum FOV of 2.4°
[d]Esposito et al. (2005a)
[e]Flasar et al. (2005b)

Table 2.22. Technical data for several *Cassini* instruments designed to study Saturn's magnetosphere and dust

Instrument acronym	Subsystem, mode or channel	Function
CAPS[a]	Electron spectrometer (ELS)	Measure electrons with kinetic energies of between 0.6 eV and 28 keV
"	Ion beam spectrometer (IBS)	Measure ions with kinetic energies of between 1 eV and 50 keV per charge
"	Ion mass spectrometer (IMS)	Measure composition of magnetospheric plasmas 1 eV to 50 keV per charge
INMS[b]	Closed system neutral mode	Detect non-reactive[c] neutral atoms at concentrations of at least 10^4/cm^3
"	Open source neutral mode	Detect reactive[c] neutral atoms at concentrations of at least 10^5/cm^3
"	Open source ion mode	Detect cations with kinetic energies less than 100 eV at concentrations of at least 10^{-2}/cm^3
MIMI[d]	Ion and neutral camera (INCA)	Detect neutral species produced from plasmas
"	Charge-energy-mass-spectrometer (CHEMS)	Determine ion mass, charge and energy (3–230 keV/nucleon)
"	Low energy magnetospheric measurement system (LEMMS)	Detect ions and electrons in both the forward and back directions
RPWS[e]	Orthogonal 10-m electric field antennas	Detect electric waves with frequencies of between 1 and 16 MHz; detect dust
"	Search coil magnetic antennas	Detect magnetic fields with frequencies of between 1 and 12 kHz
"	Langmuir probe	Measure electron densities and temperatures
CDA[f,g]	Polyvinylidene fluoride (PVDF) sensors	Determine the mass of dust particles used for rates of impact of up to 10^4 particles/second
"	Dust analyzer (DA)	Measure the mass, electrical charge, chemical composition, velocity of dust within the mass range of 5×10^{-18} to 1×10^{-12} kg

[a]Young et al. (2004)
[b]Waite et al. (2004)
[c]A non-reactive atoms is one which does not react with the instrument surface and a reactive atom is one which reacts
[d]Krimigis et al. (2004)
[e]Gurnett et al. (2004)
[f]Srama et al. (2004)
[g]Kempf et al. (2005)

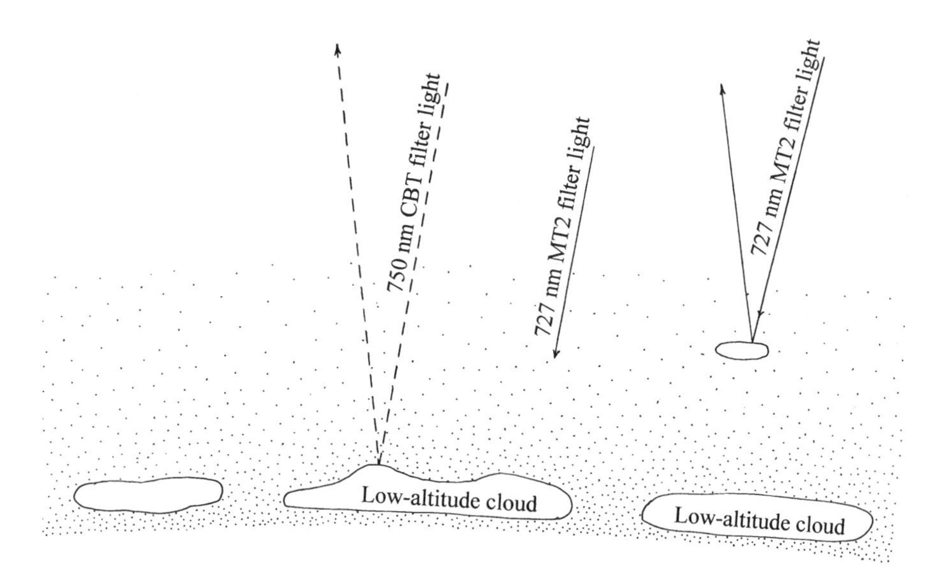

Fig. 2.31. The light which the MT2 filter transmits is unable to penetrate through much of Saturn's atmosphere because of methane absorption. The light that the CBT filter transmits, however, is able to penetrate to low altitude clouds that are at the 0.5 atm. level. This is based on information in Porco et al. (2005a) (Credit: Richard Schmude, Jr.).

Visual, Infrared and Ultraviolet Instruments

Four *Cassini* instrument packages measure ultraviolet, visible and infrared wavelengths. We'll describe relevant findings from each of these starting with the Cassini Imaging Science Subsystem (ISS).

Cassini Imaging Science Subsystem (ISS)

Have you ever noticed the stronger winds at the top of a mountain? Essentially, the wind speed changes at different altitudes. Saturn's atmosphere also has this characteristic. The Cassini Imaging Science Subsystem (ISS) imaged clouds in both visible and methane band light during 2004. Porco et al. (2005a) report peak wind speeds on Saturn dropped from 450 m/s in 1980–1981 to 360 m/s in 2004. They point out that the measured wind speeds depend on the altitude of the clouds. Let's see how this works.

The ISS contained several filters. Two of them, with wavelengths of peak transmission in parentheses, are the CB2 (750 nm) and MT2 (727 nm) filters. Since methane absorbs light with a wavelength of 727 nm, the MT2 filter detects features at high altitudes. The light that the CB2 filter transmits, however, passes through the methane. As a result, it detects features at lower altitudes. See Fig. 2.31. During 2004, the ISS imaged the same regions with both filters. Features in the CB2 filter moved 100 m/s faster than those imaged in the MT2 filter. Essentially, clouds at the lower altitude moved faster.

Fig. 2.32. The Imaging Science Subsystem (*narrow angle*) on *Cassini* recorded two images of Phoebe on June 11, 2004. Astronomers then created this mosaic from the images (Credit: NASA/JPL/Space Science Institute).

The ISS has also given us close-up images of Phoebe, one of Saturn's distant moons. See Fig. 2.32. Thanks to ISS images, we know that Phoebe's surface has a high concentration of craters and, hence, is very old. Phoebe's surface also has different layers of material. Astronomers determined its average diameter to be 213 ± 2 km and its average density to be 1630 ± 45 kg / m^3.

We have also learned more about Saturn's Moon Iapetus from the ISS. Images show a large ridge around it. See Fig. 2.33. This feature stretches almost 120° of longitude. It lies near Iapetus's equator. The ridge rises as much as 20 km (or 12 miles) above the surface. ISS images also show that the dark area on Iapetus (Cassini Regio) is heavily cratered and, hence, is very old.

The ISS also imaged Saturn's rings. Three goals of ISS ring studies were to measure the colors of individual ringlets, search for spokes and collect information on ring dynamics. Porco et al. (2005a) used ISS images made in different wavelengths to measure ring colors. From these images, they concluded that the different colors are caused by varying amounts of contaminants. This group also reports that spokes were not present from February to June of 2004. This is consistent with them being more common when the Sun-ring plane angle is below 15°. The ISS images show a variety of ring phenomena, including density waves. Astronomers were able to estimate the mass of Saturn's moons Atlas and Pan from density waves. Although the ISS collected lots of valuable data, the Composite Infrared Spectrometer (CIRS) also collected important data. Before discussing the ISS, we will explain altitudes on Saturn.

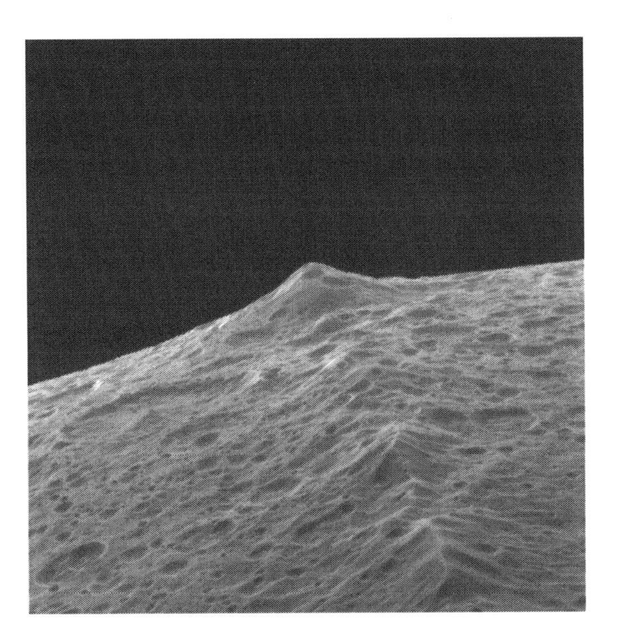

Fig. 2.33. An image of a large ridge on Iapetus. The wide-angle camera on *Cassini* took this image on September 10, 2007 (Credit: NASA/JPL/Space Science Institute).

Altitudes on Saturn

On Earth, altitudes are expressed in terms of sea level. For example, the tallest mountain in the continental United States is Mount Whitney, which is 4,418 m above sea level. Unlike Earth, Saturn lacks a solid or liquid surface. Therefore, altitudes are expressed in terms of pressure. The lower the pressure, the higher the altitude. For example, an area at the 0.01 atm. level is higher than one at the 1.0 atm. level. See Fig. 2.34.

Composite Infrared Spectrometer (CIRS)

The Composite Infrared Spectrometer (CIRS) measured the composition and temperature of Saturn's atmosphere. CIRS data show that its atmosphere contains 0.5% methane. This instrument also measured the temperature at different altitudes. Between the 0.5 and 0.1 atm. level, the temperature falls with increasing altitude; this is the same trend that occurs in our troposphere. Above the 0.03 atm. level, it increases with altitude. See Fig. 2.34. A similar trend occurs in our stratosphere. The CIRS also measured the temperatures of Saturn's moon Phoebe. The respective day and night temperatures are about 110 and about 75°K.

How does the CIRS measure temperature? Before we answer this question, let us recall what temperature is exactly. The temperature of an object is a measure of the average kinetic energy of its atoms. The higher the energy, the higher the temperature. Or, in other words, the faster that atoms move, the higher the temperature. An object at 288°K emits infrared wavelengths. If it behaves as a blackbody it should give off more radiation at a wavelength of 10 mm than at other wavelengths.

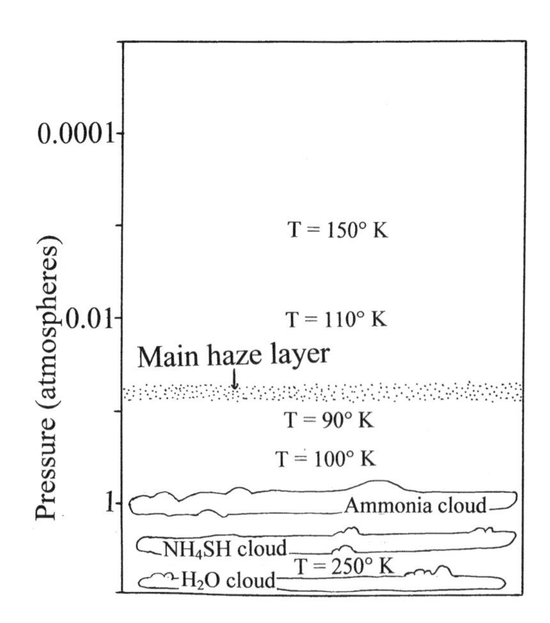

Fig. 2.34. A cross-section view of Saturn's atmosphere. As the altitude rises, the pressure drops. Representative temperatures and cloud layers are also shown. The temperatures and cloud data are from Flasar et al. (2005a) and Beatty and Chaikin (1990) (Credit: Richard Schmude, Jr.).

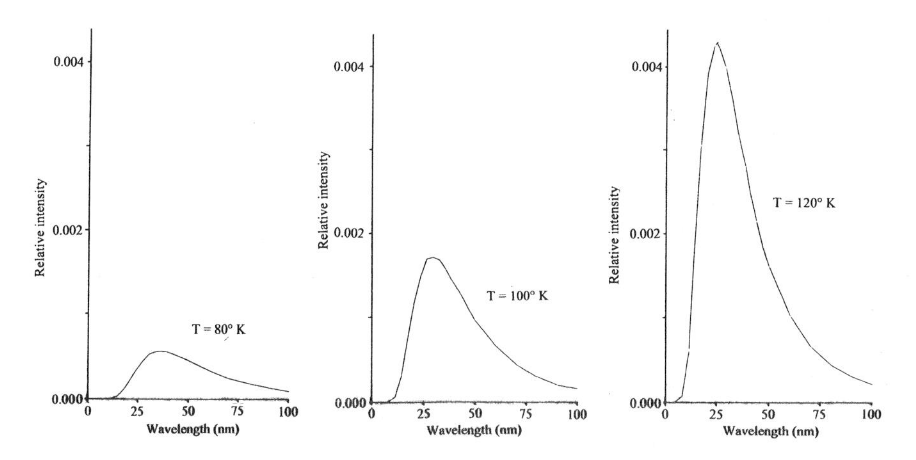

Fig. 2.35. Three graphs of the relative intensity versus wavelength for blackbodies at temperatures of 80°K, 100°K and 120°K. This author used equations from Atkins and de Paula (2002) to compute the blackbody curves (Credit: Richard Schmude, Jr.).

Figure 2.35 shows the intensity of radiation versus wavelength for blackbodies at 80°K, 100°K and 120°K. These graphs have two differences. One is the relative intensity that increases with temperature. The second difference is the wavelength at maximum intensity. It is 24 mm at 120°K but is 36 mm at 80°K.

Blackbodies at different temperatures emit different amounts of infrared radiation. Therefore, one may measure temperature from emitted light. The CIRS measures the intensity of infrared radiation at different wavelengths. Astronomers fit intensities to different blackbody temperature curves. Each temperature has a different curve. The temperature that fits the data best is the one selected. See Fig. 2.36.

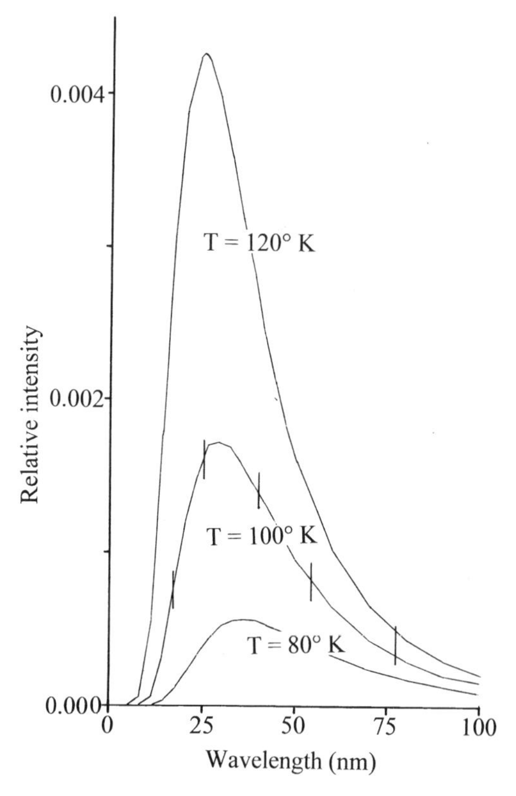

Fig. 2.36. In order to determine the temperature of a large body such as a planet, one should measure the intensities of emitted light at several different wavelengths. In this case, the intensities of mid-infrared light are shown as vertical bars. One then compares different blackbody curves at different temperatures until there is a fit. In this case, the curve at 100°K fit the data. The blackbody temperature is, therefore, 100°K. This author used equations from Atkins and de Paula (2002) to compute the blackbody curves (Credit: Richard Schmude, Jr.).

The CIRS also records infrared spectra. This enabled one group (Guerlet et al. 2009) to measure the amount of acetylene (C_2H_2), ethane (C_2H_6) and propane (C_3H_8) in Saturn's atmosphere. This group also discovered that the amount of propane increases as one approaches Saturn's south polar region. They also found an enhancement of C_2H_2 and C_2H_6 in the area of the ring shadow. Guerlet et al. (2010) report that the amounts of diacetylene (C_4H_2) and methylacetylene (CH_3C_2H) varies with the latitude. The abundance of these gases also changes with altitude. These results will serve as constraints for future models of atmospheric circulation. Like the ISS and CIRS, the Ultraviolet Imaging Spectrograph (UVIS) also carried out measurements that will be described.

Ultraviolet Imaging Spectrograph (UVIS)

When *Cassini* was 4 million km from Saturn, the UVIS recorded images of Saturn and the surrounding area. These were in the specific wavelength of light that oxygen emits. Astronomers used the intensities to determine the amount of this gas in Saturn's magnetosphere. They discovered that it dropped a factor of two over a 2-month period. Esposito et al. (2005b) suggest the rings and moons are sources of oxygen.

The ISS, CIRS and UVIS have all contributed to our understanding of the Saturn system, but the Visual and Infrared Mapping Spectrometer (VIMS) has helped us better understand the surface of Saturn's largest moon, Titan.

Visual and Infrared Mapping Spectrometer (VIMS)

The VIMS is able to image Titan's surface. This is possible because its atmosphere is nearly transparent to some wavelengths. Therefore, all one has to do is record an image at the appropriate wavelength and it will show Titan's surface.

Images made by VIMS show Titan's sand dunes. Barnes et al. (2009) used images to estimate a dune spacing of 2.1 km. This group also reports dunes are oriented in a nearly east–west direction. They report one dune is between 30 and 70 m tall. These results are consistent with radar and ISS results.

In addition to sand dunes, the VIMS helped us learn more about a lake on Titan. Brown et al. (2008) summarizes images and spectra of Ontario Lacus. This is a lake about the size of Lake Ontario. Astronomers believe it is a body of liquid ethane mixed with other substances.

Magnetometers

The magnetic field of a planet (or moon) yields a wealth of information about the interior. For example, the magnetometers on *Galileo* detected a magnetic field near Jupiter's moon Europa (The Worlds of Galileo ©2001 by Michael Hanlon, p. 140.). Furthermore, the data showed that Europa's magnetic field changed in a short period of time. Some scientists believe this is evidence for a liquid ocean below Europa's icy surface. Could one or more of Saturn's moons have a liquid ocean below its surface? Is Saturn's magnetic field changing direction?

These and other questions are what motivated scientists to develop the magnetometer assembly on *Cassini.* One problem with a magnetometer, though, is that electrical equipment generates magnetic fields. This is a problem since *Cassini* is chock full of electrical equipment.

To overcome this problem, three courses of action were undertaken. First, the magnetometers were placed on a long pole called a boom. Therefore, they are several meters from the electrical equipment. Secondly, a special committee called the Magnetic Control Board was set up. It was made up of scientists and its purpose was to insure that spacecraft-generated magnetic fields did not interfere with magnetometer readings. Thirdly, scientists designed their instruments so they did not produce magnetic fields over the limit set by the Magnetic Control Board. The Magnetic Control Board wanted *Cassini's* magnetic field to be lower than 0.2 nT at the outermost magnetometer.

The magnetometers are able to measure both the strength and direction of the magnetic field. They may also measure fields stronger than 10,000 nanoTesla (nT) with a resolution of about one nT. For a comparison, Earth's magnetic field at the surface is between 30,000 and 60,000 nT.

The magnetic field depends on Saturn's interior, the magnetosphere and interactions between the solar wind and the magnetosphere. The magnetometers made hundreds of measurements in 2004. Dougherty et al. (2005) report the strength and orientation of Saturn's magnetic field did not change between 1981 and 2004. They report, however, the magnetosphere-generated field changed during this time.

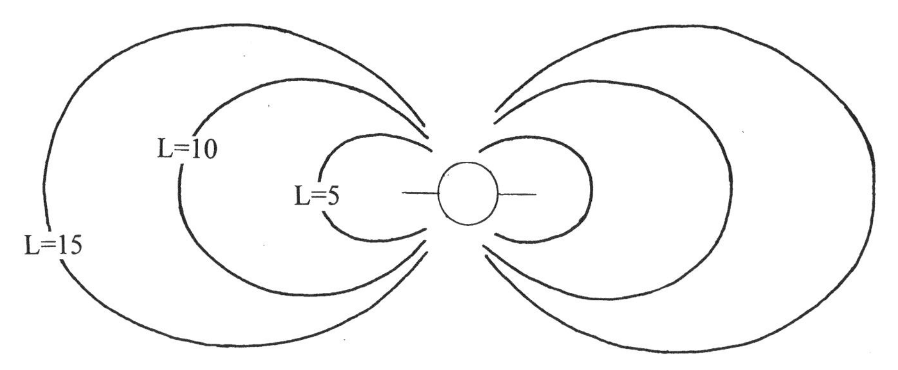

Fig. 2.37. Lines of magnetic force (L) associated with Saturn's magnetic field are illustrated. The lower the L value, the stronger is the magnetic field (Credit: Richard Schmude, Jr.).

Instruments That Study Saturn's Magnetosphere

Saturn's magnetic field extends far into space. It has invisible lines of force. See Fig. 2.37. Essentially, as the L value drops, the line of force gets stronger. The field does not affect the movement of neutral atoms (those with no electrical charge). It also does not affect the moons and rings. However, it affects charged particles such as ions and electrons.

Saturn's magnetosphere is the region where the magnetic field controls the movement of charged particles. This region extends over a million kilometers from Saturn. At the outer boundary of the magnetosphere, the planetary magnetic field becomes weaker than that in the solar wind. Ions, plasmas and fast-moving electrons are in Saturn's magnetosphere. (Plasma is an electrically neutral mixture of ions, electrons and neutral gas.) The distribution of this material depends on the magnetic field, dust, rings and moons.

Cassini has four instrument packages that gather information on the magnetosphere. Table 2.22 summarizes them. Some contain more than one instrument (these are labeled a subsystem in the table). Others operate in more than one mode or channel. The main function of each subsystem, mode or channel is listed.

The Ion and Neutral Mass Spectrometer (MIMI) measures the characteristics of ions and electrons in Saturn's magnetosphere. One important MIMI finding is that nitrogen ions (N^+) are scarce between the $L = 5.7$ and $L = 12.6$ shells. This is consistent with little or no nitrogen escaping from Titan. Figure 2.37 describes L shells. MIMI data also show that ions are scarce near the rings and larger moons. Essentially, these objects are absorbing ions. In addition to the MIMI instrument, the Ion and Neutral Mass Spectrometer (INMS) also collects data on the magnetosphere.

The INMS measures the mass-to-charge ratio of several species. This is an ion's mass, in atomic mass units or amu, divided by its charge. For example, H_2O^+ has a mass of 18 amu and a charge of $+1$ and, hence, its mass-to-charge ratio is 18. The mass-to-charge ratio of H_2O^{+2}, however, equals $18 \div 2 = 9$. In 2004, one group of astronomers reported that the INMS detected mass-to-charge peaks at 8, 16, 17, 18, 23, and 32. The peaks at 16 and 32 are probably the oxygen ions O^+ and O_2^+.

A third instrument that collects data on Saturn's magnetosphere is the Radio and Plasma Science (RPWS) instrument. It measures radio waves with frequencies

Scientific Satellite Spacecraft

between 50 and 500 kHz. These are believed to change as Saturn's magnetic field rotates and are called Saturn Kilometric Radiation (SKR). During 1980 and 1981, the SKR changed every 10.656 h. Astronomers selected this as Saturn's rotational period. In 2004, the SKR changed every 10.763 h. This is a difference of just over 0.1 h or 6 min, which is larger than the uncertainty of the measurements. One possibility is that Saturn's rotation rate has changed 6 min. Alternatively, something else is going on. More measurements are needed to understand this change.

The RPWS instrument detects Saturn Electrostatic Discharges (SEDs). These are bursts of radio waves with frequencies of between 100 and 40 MHz. Astronomers believe lightning is the cause of SEDs. The *Voyager* spacecraft first detected these events in 1980 and 1981. *Cassini's* RPWS instrument will enable astronomers to monitor the frequency of SEDs over several years.

As *Cassini* approached Saturn, the RPWS instrument detected bursts of SEDs. Through careful study, astronomers discovered three sets of events had periods of 10.72, 10.26 and 10.66 h. They matched two sets of SEDs with specific clouds.

Imagine a 100-mile-long lightning bolt. The RPWS data are consistent with this size of bolt on Saturn. Essentially, Saturn's strongest lightning bolts have about a million times the energy of those on Earth. Although the RPWS collected valuable data, the Cassini Plasma Spectrometer (CAPS) was also useful.

The CAPS carried out a survey of Saturn's magnetosphere. One group (Young et al. 2004) report that the CAPS data is consistent with Saturn's magnetosphere having four different parts. The outer region contains plasma made up mostly of protons. A second inner layer contains protons mixed with the ions O^+, OH^-, H_2O^+ and H_3O^+. A third inner layer rotates at nearly the same speed as Saturn and has lots of O^+, OH^-, H_2O^+ and H_3O^+ ions. The fourth layer lies near the A ring and has lots of O^+ and O_2^+ ions. Astronomers believe the high concentration of O^+ and O_2^+ near the rings is evidence of some oxygen lying above them. Young et al. (2004) estimate a quantity of this gas consistent with it being 10^{-12} times that of Earth's atmosphere at sea level. They suggest that high energy particles collide with the ice particles in the rings. This then causes the oxygen to escape.

The CAPS instrument also detected nitrogen ions (N^+) deep in the magnetosphere. These had low kinetic energies, which led one group to believe that the nitrogen is ionized in the magnetosphere. Much of this gas may have escaped from Titan's atmosphere.

Dust Detecting Equipment

Dust is very common in our Solar System. It causes the zodiacal light and the gegenschein. Two dust sources near Saturn are the rings and moons. One question that astronomers want an answer to is, what is its chemical composition, charge, size and concentration of dust? They are also interested in learning more about the ice plumes above Saturn's moon Enceladus.

The Cosmic Dust Analyzer (CDA) and the Radio and Plasma Wave Science investigation (RPWS) are the two dust detectors on *Cassini.* The CDA collects data on the mass, electric charge, chemical composition and velocity (speed and flight direction) of dust near Saturn. It contains a time-of-flight mass spectrometer that measures the chemical composition of ions produced from particle collisions. The RPWS, on the other hand, contains three 10-m electrical antennas that are at nearly

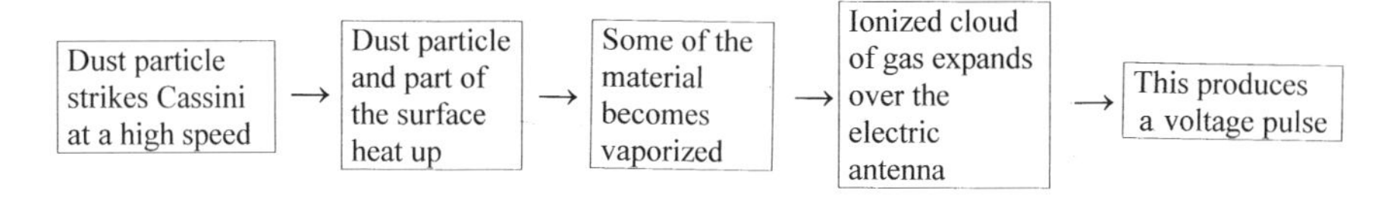

Fig. 2.38. The sequence of events that occur when a dust particle collides with one of the antennae on *Cassini*. See Gurnett et al. (2005) (Credit: Richard Schmude, Jr.).

right angles from one another. When a particle collides with an antenna, it is vaporized, and part of the vapor becomes ionized. The resulting ions create an electrical signal the antenna detects. Scientists compute the particle mass by measuring the signal strength. The detection process is illustrated in Fig. 2.38.

In 2004, astronomers worried about contamination from *Cassini's* exhaust. About 850 lb of exhaust was released during the orbital insertion. Fortunately, the combustion products did not affect the initial CDA measurements.

Radio Equipment

The *Cassini* Titan Radar Mapper (RADAR) uses radar to study the Saturn system. This instrument sends out short radar pulses. The RADAR instrument or the Deep Space Network detects them. This instrument also reconstructs radar images. Here is how it works. Radar with a frequency of 13.8 GHz (wavelength = 2.17 cm) is aimed at Titan. It bounces off the surface and back to the receiver. The computer constructs an image. The image contains pixels just like a CCD image. The brightness of each pixel is proportional to the amount of radar reflected. The data from the RADAR instrument enabled astronomers to learn more about Saturn and its largest moon, Titan. For example, the RADAR data show evidence of evaporation on Titan. Essentially, some of the lakes appear to have lost liquid. Some images even suggest dried up lakes. These observations have led one group of astronomers (Stofan et al. 2007) to suggest Titan's lakes may dry up in the summer and grow in the winter.

Astronomers used radar to measure the mass of two of Saturn's moons. Let's see how they did this. When a detector moves towards or away from a source of electromagnetic radiation, a small shift in frequency occurs. This is called the Doppler effect. See Chap. 1. By measuring this shift, astronomers measure velocity changes. This in turn yields a value of the gravitational acceleration of a nearby moon. With the acceleration and Newton's law of universal gravitation, astronomers compute the moon's mass.

Astronomers used the Doppler effect to measure the mass and density of Saturn's moons Phoebe and Iapetus. For example, Rappaport et al. (2005) report masses of 8.27×10^{18} kg for Phoebe and 1.80×10^{21} kg for Iapetus. Our Moon has about forty times the mass of Iapetus. By combining the diameter and mass, this group also reports densities of $1,600\pm20\ \text{kg}/\text{m}^3$ for Phoebe and $1,105\pm\ 27\ \text{kg}/\text{m}^3$ for Iapetus. These values are larger than pure water ice ~940 kg/m^3. This is evidence that both contain other materials besides ice.

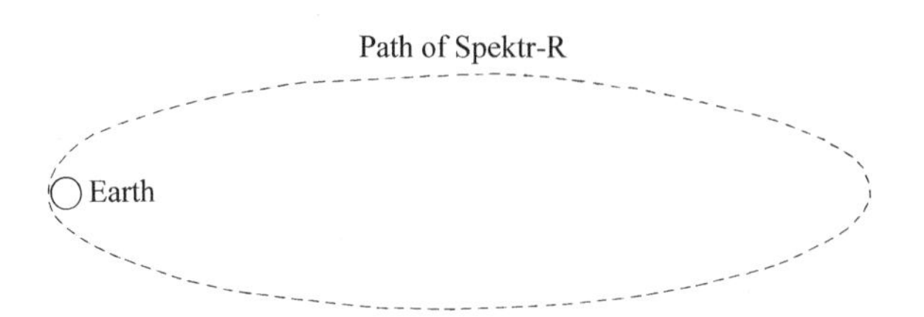

Fig. 2.39. The orbit of the Russian radio telescope *Spektr-R*. Its altitude will range from 500 to 340,000 km (Credit: Richard Schmude, Jr.).

Spektr - R

Russia launched its Spektr-R radio telescope in July 2011 (Clery 2011). It follows a highly elliptical orbit around Earth. See Fig. 2.39. This satellite has a 10-m radio dish. When combined with ground-based telescopes it is expected to produce radio images of unprecedented resolution. This is how it will work. Astronomers will use a technique called interferometry. Essentially, they will combine the signals from two or more telescopes at different locations. When this is done, the resulting signal is equivalent to that from a single instrument equal to the spacing between the telescopes. For example, the Very Large Array (VLA) works on this principle. Essentially, it is 27 radio telescopes arranged in a Y pattern. These may be up to 40 km (25 miles) apart. With this arrangement, VLA signals are equivalent to those coming from a single 40 km instrument. Astronomers may attain a radio-frequency resolution of ~1.0 arc second from this facility.

As a second example, the Very Long Baseline Array or VLBA is a group of radio telescopes spread across the United States that are linked to each other. Signals from the VLBA are equivalent to those from an instrument the size of the United States. The motivating factor behind both the VLA and VLBA is the larger the telescope, the higher the resolution. Russian astronomers plan to combine data from Spektr-R and Earth-based telescopes to generate signals that would be equivalent to an instrument around 300,000 km across. They are hoping to attain radio-frequency resolutions better than 0.001 arc second.

Gaia

Between 1989 and 1993, the *Hipparcos* satellite measured positions of about one million stars. It revolutionized our view of our galaxy and the universe. For example, it has helped refine the distance scale for Cepheid variable stars. Henrietta Swan Leavitt (1868–1921) first discovered the period-luminosity relationship of these stars in 1908, and since then astronomers have used them to determine the distance to galaxies and other objects.

One byproduct of the *Hipparcos* data is the Millennium Star Atlas ©1997 by R. W. Sinnott and M. A. C. Perryman. This atlas shows the positions of about a million stars down to magnitude 12 on a convenient grid, the proper motion of thousands of stars, positions and orientations of thousands of deep sky objects, positions and separations of about 22,000 double stars and approximate brightness

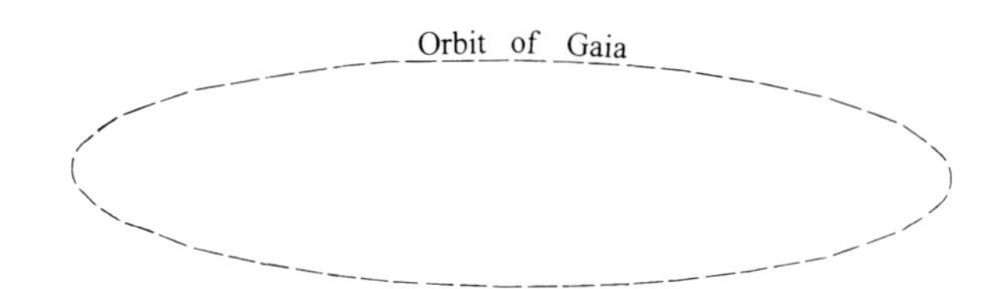

Fig. 2.40. *Gaia* follows an elliptical path around the Sun-Earth L_2 point (Credit: Richard Schmude, Jr.).

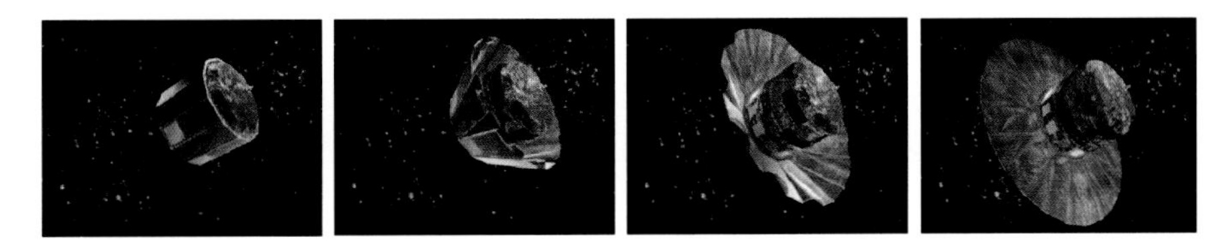

Fig. 2.41. This sequence shows how *Gaia's* Sun shield will open up (Credit: ESA – C. Carreau).

ranges for variable stars. This atlas also shows the brightness of each star in about the same way that we would see it in the sky. The follow-up spacecraft of *Hipparcos* is *Gaia*. Most of the technical details which are presented about *Gaia* are from the websites http://sci.esa.int/science-e/www/area/index.cfm?fareaid=26, http://gaia.esa.int/, from Dambeck (2008) and McDowell (2006h).

The European and Russian space agencies plan to launch this spacecraft in 2013 with a Soyuz rocket from French Guiana. It will travel to near the Sun-Earth L_2 point and enter a Lassajous-type orbit. Figure 2.40 shows the orbit and Figure 1.25 shows the L_2 point.

Gaia will move in an elliptical path around the L_2 point. The orbital period around L_2 will be 180 days. The advantage of such an orbit is that the spacecraft will avoid entering Earth's shadow. Earth's shadow will lead to significant cooling of the instruments and will introduce error. Since the L_2 point orbits the Sun once a year, this spacecraft will image all parts of the sky. Plans call for *Gaia* to spin slowly around an axis pointed 45° from the Sun. This will allow the solar arrays to collect an adequate amount of sunlight. It will also insure transmitters are aimed at Earth. Radio antennas located in Cebreros, Spain, and New Norcia, Australia, will receive *Gaia's* signals.

Once *Gaia* is in deep space, its Sun shield will unfold. See Fig. 2.41. This shield will serve two purposes. First, it will supply power. Second, it will prevent direct sunlight from heating the instruments. In space nothing spreads out direct sunlight. Therefore, components in direct sunlight may reach high temperatures. The Sun shield will keep the electrical components in continuous shade, and this will lead to a nearly constant temperature.

The mirror-camera assembly will carry out the critical measurements. It will contain two rectangular mirrors with dimensions of 0.5 by 1.45 m (20 in. by 57 in.). These will collect starlight and focus it onto optical components. *Gaia* will image two areas of the sky simultaneously. These will be 106.5° apart. Each image is expected to contain around 20,000 stars. Throughout its 5-year mission, each object brighter than magnitude 20 should be imaged about 70 times. These objects will include stars, asteroids, brown dwarfs, comets, planets and natural satellites. *Gaia* should also image artificial satellites. Table 2.23 lists some technical details of *Gaia*.

Table 2.23. A few technical details of *Gaia*

Characteristic	Value
Mass (at launch)	2030 kg[a]
Project duration	2013–2018[a]
Spin rate	One complete revolution every 6 h[b]
Orbital period around L_2	180 days[a]
Orbital period around our Sun	1 year[a]
Field-of-view	0.7 degrees[c]
Temperature of electrical components	−100°C (or −148°F)[a]
Spectral range	320–1000 nm[a]
Accuracy or star position measurement	0.000024 arc – seconds[c,d]

[a]From http://sci.esa.int/science-e/www/area/index.cfm?fareaid=26
[b]Dambeck (2008)
[c]From McDowell (2006h)
[d]For stars brighter than magnitude 15, measurements will be less accurate for fainter stars

Gaia will carry out three types of measurements. Firstly, it will measure positions. Objects brighter than about magnitude 15 will be measured to an accuracy of 24 micro arc seconds. This is equivalent to a flashlight at the Moon's distance moving 4.6 cm. Secondly, this spacecraft will carry out brightness measurements. Since CCD detectors will be used above our atmosphere, the accuracy of the measurements should be around 0.001 magnitudes for objects brighter than magnitude 15 and may be 0.02 magnitudes for objects at magnitude 20. Thirdly, it will record spectra. According to McDowell (2008h), it will use the spectra of 150 million stars to determine radial velocities. With these measurements, astronomers will be able to determine both the velocity of stars in all three dimensions and their distances. Let's explain.

Gaia will start off imaging a star's spectrum. Let's assume it images an absorption spectrum showing the hydrogen-alpha line. If the star is not moving towards or away from us, this line should be at a wavelength of 656.3 nm. If, however, this star is moving towards us the hydrogen-alpha line will shift to a shorter wavelength. The hydrogen-alpha absorption line should shift to a longer wavelength, however, if the star is moving away from us. See Fig. 2.42. This shift is caused by the Doppler effect, which is described in Chap. 1. This shift enables astronomers to measure the velocity towards or away from us. Accurate position measurements enable astronomers to measure the other two velocity components. Therefore, by measuring the Doppler shift along with precise star positions, astronomers can measure the velocity of each star in all three dimensions.

Gaia will also measure the parallax of stars. The parallax of an object is the apparent shift in its position caused by the changing line of sight. The line of sight runs through the observer's eye (or instrument) and the object. In order to illustrate parallax, try the following experiment. Focus on a nearby object with a more distant object in the background. Open just your left eye and note the position of the object with respect to the background object. Now close your left eye and open your right eye. The nearby object should have shifted position with respect to the background object. Now view the object by opening one eye and then the other. The object should appear to move back and forth with respect to the background object. Repeat this experiment for a more distant object. The more distant object may appear to shift, but the shift should be smaller than what it was

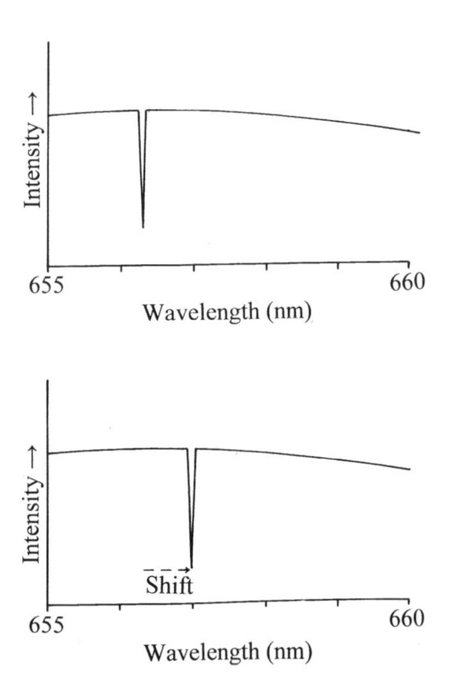

Fig. 2.42. The hydrogen absorption feature is at 656.3 nm when the star is not moving towards or away from the observer (*top spectrum*). If the star is moving away from the observer, the absorption feature will shift to a longer wavelength (657.0 nm in the *bottom spectrum*). This is the Doppler effect (Credit: Richard Schmude, Jr.).

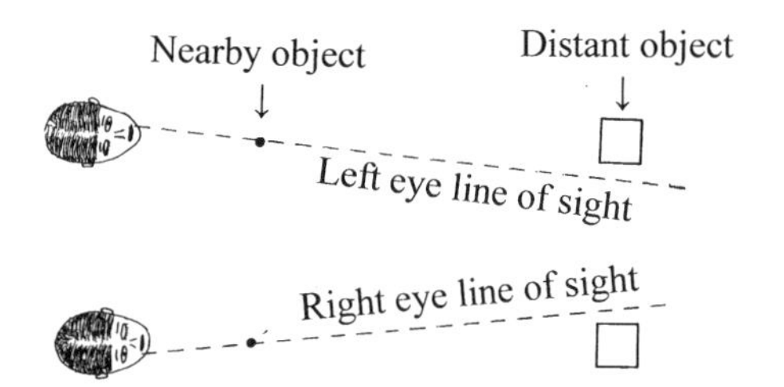

Fig. 2.43. A nearby object shifts position when one looks at it with just the left eye and then with just the right eye. The nearby object appears to be to the *right* of the distant object when viewed with just the left eye. The opposite occurs through the right eye. This shift is caused by the changing line of sight and is called parallax (Credit: Richard Schmude, Jr.).

for the nearby one. The right and left eye are about 7 cm apart, and therefore, the line of sight changes. This change is large for nearby objects. It becomes small, however, for more distant ones. See Fig. 2.43.

As *Gaia* moves around our Sun, its line of sight changes. The stellar parallax is the shift in the position of a star. In our case, *Gaia* is the observer. For *Gaia*, the relationship between parallax (p) and distance (d) is:

$$d = \frac{1.0\,\text{arc} - \text{seconds} \times \text{parsec}}{p} \tag{2.1}$$

Therefore, if a parallax of 24 micro arc seconds (0.000024 arc – seconds) is measured, the distance would be just over 42,000 parsecs. This is near the edge of our galaxy.

One difficulty with measuring the parallax of a star is the distance of the background stars. If these stars are almost as close as the one being measured the distance would be incorrect. To make matters more complicated, we are dealing with a large number of stars. It is for these reasons that *Gaia* will solve for stellar distances differently. At the end of its mission, it will help astronomers solve for the distances of at least 150 million stars all at once. This was impossible with 1993 computer technology, but it should be possible in 2018. The star data from *Gaia* may be available as early as 2020.

How will astronomers use *Gaia* data? They plan to use it to focus on several areas, including our Solar System, neighboring stars, our galaxy, objects beyond our galaxy and tests on Einstein's General Theory of Relativity. Here you will learn about a few of these.

Solar System Studies

Gaia is expected to image about 250,000 asteroids. Centaurs and Trans-Neptune objects (TNOs) will also be targets. Asteroids, centaurs and most TNOs are too small to be classified as planets or dwarf planets. They orbit our Sun and are not comets. Asteroids lie inside Jupiter's orbit, Centaurs lie between the orbits of Jupiter and Neptune and TNOs lie beyond Neptune's orbit. See Fig. 2.44.

Astronomers will refine the orbits, rotational rates, photometric constants, and light curves of many of these objects with *Gaia* data. They also expect to discover several thousand new asteroids along with perhaps a few dozen centaurs and TNOs. Astronomers will undoubtedly discover new comets in *Gaia* data. In addition to this, the Trojan asteroids around Jupiter and perhaps other planets will be examined. Trojan asteroids for Jupiter are those which lie near the Jupiter-Sun L_4 and L_5 points. See Fig. 2.45.

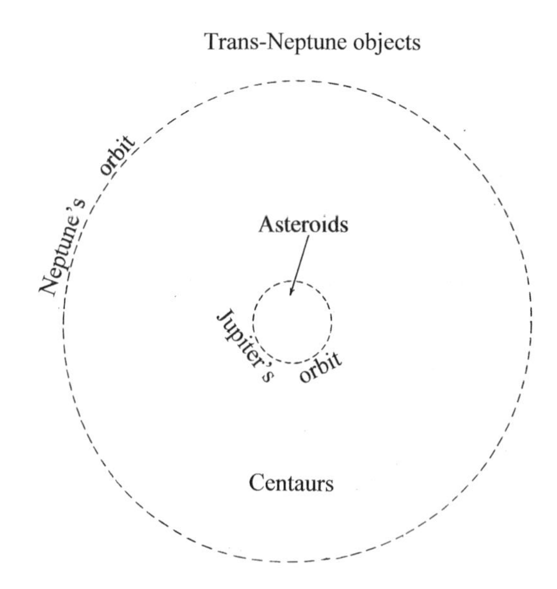

Fig. 2.44. Asteroids lie inside Jupiter's orbit, Centaurs lie between the orbits of Jupiter and Neptune and Trans-Neptune Objects lie beyond the orbit of Neptune (Credit: Richard Schmude, Jr.).

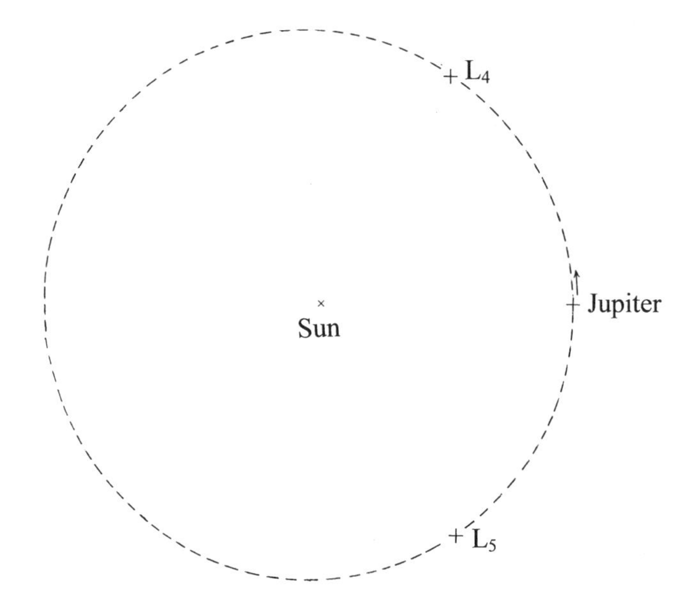

Fig. 2.45. The Trojan asteroids lie at the Jupiter-Sun L_4 and L_5 points. The L_4 point precedes Jupiter by 60° and the L_5 point follows Jupiter by 60° (Credit: Richard Schmude, Jr.).

Astronomers will use the data to determine whether Jupiter's Trojans formed near that planet or formed in another part of our Solar System and were later captured. *Gaia* will also search for objects in our Solar System that have high orbital inclinations. Astronomers hope to measure asteroid masses from *Gaia* data. Let's see how this will be done.

When two asteroids get close, they will exert a gravitational force on one another. The force that each exerts will alter the acceleration of the other. The closer they pass, the greater will be this change. In most cases, one asteroid will be larger than the other. In these cases, the smaller one will experience a larger acceleration than the other one. It is this change that *Gaia* will measure. Astronomers will then use it to compute the mass of the larger asteroid.

As of 2011, the eight major planets in our Solar System have 168 known moons. Pluto has four more. *Gaia* will gather additional positional and photometric data on many of these. This will be more accurate than similar measurements from Earth. Scientists may determine the rotational periods of the smaller moons from *Gaia* data.

One important question is "Are we seeing more long-period comets than in the distant past?" One way to answer this question is to examine the velocities and distances of nearby stars. Astronomers believe that when a star makes a close pass to our Solar System some of the icy objects are pushed towards our Sun and become comets. *Gaia* will carry out extensive measurements of all stars within 50 parsecs (or about 160 light years) of our Solar System. With this information, astronomers should be able to reconstruct more reliable star distances in the past. They should also be able to determine when stars made close encounters with the Oort Cloud. This should enable them to determine the frequency of long-period comets in the past and future.

Studies of Neighboring Stars

Astronomers plan to carry out star studies with *Gaia* data. For example, brightness measurements will be more accurate than what we have today. They hope to search for new variable stars. They may determine the percentages of different classes of variable stars. Finally, *Gaia* data will allow them to search for extrasolar planets, through the transit method, the star-wobble method or the direct imaging method. Let's look at the last two.

Consider a star having 20 times the mass of a nearby planet and both objects are 20 light years from Earth. Furthermore, let's assume no other planets are orbiting the star and the planet makes one trip around the star in 2 years. As it turns out, the planet and star move around the center of mass (CM) of the planet-star system. In our case, the CM lies a short distance from the center of the star. The planet travels a greater distance than the star because it is farther from the CM. Nevertheless, the star will also move. Examine Fig. 2.46a. When the planet is to the right of the CM, the star is to the left of it. One year later (Fig. 2.46b) the planet and star have shifted. They are now left and right of the CM, respectively. Therefore, in 1 year, the star has shifted from one side of the CM to the other. If the planet is 1.0 astronomical unit (au) from the star the planet and star would have shifted 1.9 and 0.1 au, respectively. At a distance of 20 light years, the star will shift 0.03 arc seconds. *Gaia* will be able to detect the shift. Astronomers hope to discover thousands of new extrasolar planets using this technique.

Will *Gaia* be able to image planets around other stars? Possibly; let's explain. If we assume the planet in Fig. 2.46 has the same albedo and size as Jupiter it would reach a peak brightness of magnitude 21. This is below *Gaia's* limit. To make matters worse, at this peak brightness, the planet will be near superior conjunction and, hence, close to the star. See Fig. 2.47. At other times the planet will be farther from the star but will be dimmer because of its smaller phase. Some extrasolar planets, however, may have extensive ring systems. *Gaia* could detect such a planet. In addition to this, several stars are closer to us than 20 light years. Perhaps *Gaia* will image a planet around one of these.

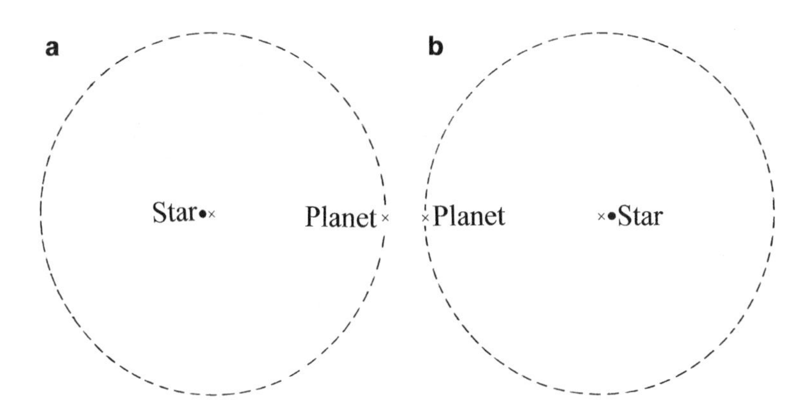

Fig. 2.46. The × is the *center* of mass of the planet-star system. When the planet is to the *right* of the ×, the star is just a little to the *left* (frame a). One half of a revolution later, the planet is now *left* of the ×. During this time the star has also moved around the *center* of mass and is just to the *right* of the × (frame b). In some cases, astronomers will be able to detect the shift in star position and, hence, determine the mass of the planet (Credit: Richard Schmude, Jr.).

Scientific Satellite Spacecraft

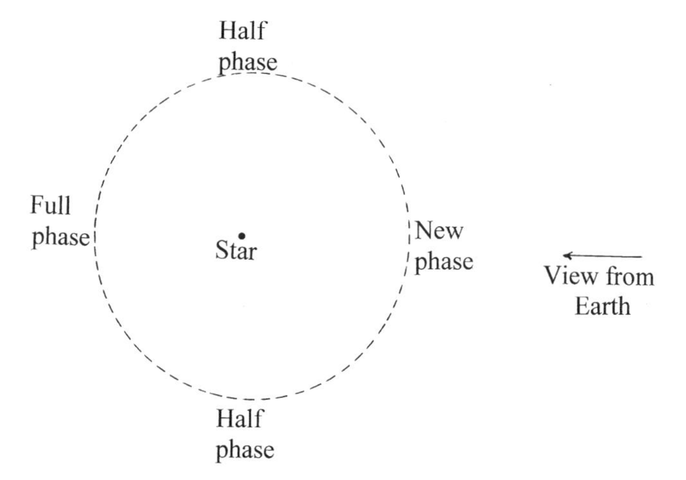

Fig. 2.47. An extrasolar planet will be at peak brightness at full phase. The problem though is it will also appear next to the star as seen from Earth. An extrasolar planet will be easiest to image when it is farthest from its parent star. In this situation, it will be at half-phase and, hence, dimmer than at full phase (Credit: Richard Schmude, Jr.).

Studies of Our Galaxy

Astronomers also hope to learn more about our galaxy by examining *Gaia* data. For example, they hope to better understand the ongoing collision between The Milky Way and the Sagittarius dwarf galaxy. Astronomers will search for groups of stars passing through our galaxy. In Fig. 2.48, the velocity of many stars have been plotted. The velocities may all look random, but take a closer look. Notice many of the smallest stars do not have a random motion but are instead traveling in nearly the same direction. Essentially, they are a group of stars passing through.

Objects Beyond Our Galaxy

Astronomers also hope to learn more about certain objects lying beyond our galaxy. Data from *Gaia* should lead to the discovery of many new novas and supernovas. (A nova is a mild stellar explosion usually taking place in the outer layers of a star whereas a supernova is larger explosion usually taking place near the center of a star.) Once a nova or supernova is discovered, astronomers will be alerted and will carry out follow-up measurements. They also hope to discover and characterize many new quasars. (A quasar is a distant object less than one light year across that gives off about 10^{11} times as much light as what our Sun gives off.)

Tests of Einstein's General Theory of Relativity

Astronomers plan to test Einstein's General Theory of Relativity. Essentially, they will measure star positions near our Sun. The General Theory of Relativity predicts the path of light changes when it is close to a large source of gravity. If this is

Fig. 2.48. Each dot represents a star and each *arrow* represents its velocity. Note the smallest stars are moving in nearly the same direction. This is evidence they are all part of a group (Credit: Richard Schmude, Jr.).

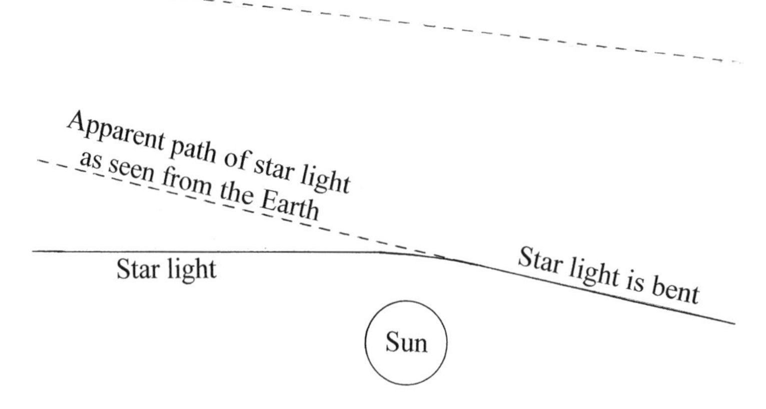

Fig. 2.49. When starlight passes a massive object, such as our Sun, gravity bends it. This bending of starlight will cause the star's position to appear to shift in the direction of the dashed line (Credit: Richard Schmude, Jr.).

correct, star positions would shift as the Sun approaches the line of sight. See Fig. 2.49.

This experiment was first carried out in the early twentieth century, but it will be verified to a much higher degree of accuracy with *Gaia* data.

Chapter 3

Earth Observing and Weather Satellites

Introduction

In 2011, dozens of satellites are collecting measurements on the atmosphere, continents, oceans, ice caps and interior. Government agencies such as the U. S. Geological Survey (USGS) operate many of them. In many cases, satellite data is sold to companies that analyze it, construct new data products and sell these. A good example is a weather company. It buys satellite data, uses it to make detailed weather forecasts, and then sells these forecasts to airlines. An airline, such as Delta, then uses the forecasts to determine whether it is safe to fly an aircraft or what areas will have extreme air turbulence. A flight might also be diverted to another airport because of extreme wind shear. Wind shear is able to tear a plane apart and, hence, this diversion can save lives. Wind shear is described in Chap. 1.

Organizations in the United States, Russia, Japan and Europe own most of the Earth observing satellites. Other countries, however, also own and operate them. Table 3.1 lists some Earth-observing satellites. Most of the information in this chapter is from the websites listed in this table. Most of the satellites operate in their own unique way and study specific characteristics of our planet. In the next few pages, you will learn about these satellites.

Earth Observing Satellites

SeaWiFS

Have you ever seen greenish pond scum? The organisms have this color because they contain chlorophyll – a green compound used in photosynthesis. Similar organisms called phytoplankton live near the surface of the ocean. They cause parts of the ocean to change color. Essentially, the more phytoplankton there is, the greener the ocean. They also affect our atmosphere by consuming carbon dioxide. Therefore, scientists interested in how these organisms impact our atmosphere

R. Schmude, Jr., *Artificial Satellites and How to Observe Them*, Astronomers' Observing Guides,
DOI 10.1007/978-1-4614-3915-8_3, © Springer Science+Business Media New York 2012

Table 3.1. Earth-observing satellites

Satellite	Year launched	Orbit Type[a]	Altitude (km)	Inclination (degrees)	Minimum ground track repeat cycle (days)	Source
SeaWiFS	1997	LELE-SS	705	98.3	1	b
Landsat-7	1999	LELE-SS	705	98.2	16	b
QuikSCAT	1999	LELE-SS	803	98.6	–	b
Terra	1999	LELE-SS	705	98.1	16	b
ACRIMSAT	1999	LELE-SS	692–713	98	–	b
EO-1	2000	LELE-SS	705	98.2	16	h
Jason-1	2001	LELE	1336	66	10	b
Aqua	2002	LELE-SS	705	98.2	16	b
GRACE	2002	LELE	485	89.5	Non-repeat	d, f
ICESat	2003	–	600	94	91	b
SORCE	2003	LELE	630	40	–	b
Aura	2004	LELE-SS	705	98.2	16	b
CALIPSO	2006	LELE-SS	705	98.2	16	g
Cloudsat	2006	LELE-SS	705	98.2	16	c
Jason-2	2008	LELE	1336	66	10	e
OSTM	2008	LE	1336	66	10	b

[a] *LE* low eccentricity, *LELE* low-Earth low eccentricity, *LELE-SS* low-Earth, low eccentricity Sun synchronous
[b] http://eospso.gsfc.nasa.gov (select *mission profiles* and then *EOS*)
[c] http://cloudsat.atmos.colostate.edu/home
[d] http://www.csr.utexas.edu/grace/
[e] http://sealevel.jpl.nasa.gov/missions/ostmjason2/
[f] Tapley et al. (2004)
[g] http://smsc.cnes.fr/CALIPSO/
[h] http://edcsns17.cr.usgs.gov/eo1/

will want to know their seasonal behavior and the overall health of large colonies. This is where the SeaWiFS (or Orbview) satellite comes in.

One of its objectives is to record accurate measurements of the ocean's color. It accomplishes this with its main scientific instrument – the sea-viewing wide field-of-view sensor. This instrument has eight filters which transmit different wavelengths of visible and near-infrared light. Filter transmission ranges are shown in Fig. 3.1. Color data enables the scientist to determine the amount of phytoplankton in the ocean. This microscopic life form is near the bottom of the food chain. Therefore, ocean ecosystems depend on it. Scientists are also interested in learning the seasonal cycle of phytoplankton, and this is best accomplished with *SeaWiFS* data.

The *SeaWiFS* satellite is able to image 90% of the ocean in just 2 days. It can resolve features as small as 1.1 km at the nadir position. (This is the position looking straight down.) The satellite is able to image an area up to 1,500 km across. This is wider than the state of Texas.

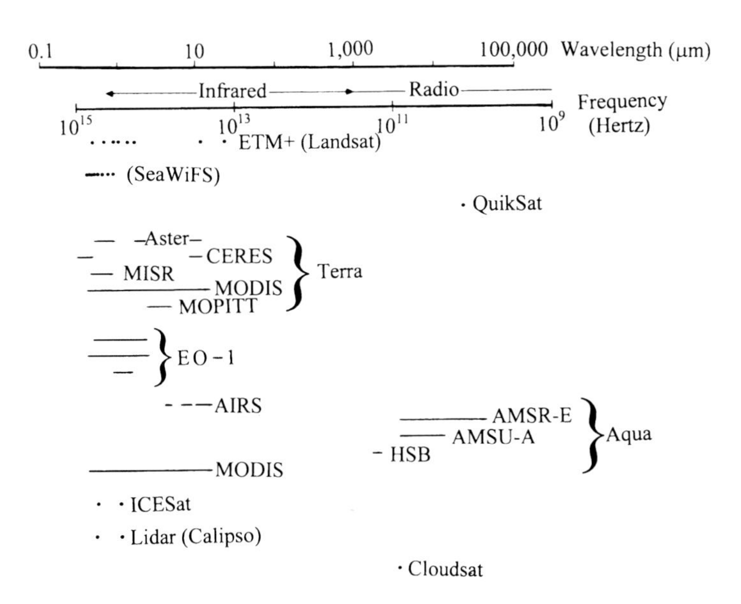

Fig. 3.1. Frequency range of several Earth-observing satellites (Credit: Richard Schmude, Jr.).

The amount of phytoplankton will affect the amount of carbon dioxide in the atmosphere. This is because the organisms consume carbon dioxide in photosynthesis. Scientists will use *SeaWiFS* data to better understand the Earth's carbon cycle and how phytoplankton impacts our atmosphere.

Landsat 7

Landsat 7 is the seventh satellite in the Landsat series dating back to 1972. Since 1972, scientists have assembled an extensive databank of Earth images. These are in different wavelengths of visible and infrared light. The objectives of *Landsat 7* are: (1) image the surface of Earth, (2) provide continuity with previous Landsat satellites, (3) offer a 16-day repetitive Earth coverage, and (4) make the images widely available at a nominal cost. Since 1999, *Landsat 7* has remained about 700 km above Earth's surface in a nearly circular orbit.

The Enhanced Thermatic Mapper Plus (ETM+) on *Landsat 7* is capable of carrying out imaging spectroscopy. It points straight down (the nadir direction). This instrument has a Ritchey-Cretien telescope with a 0.41 m primary mirror. The focal length is 2.4 m. It is coated with silver. The telescope collects light and then focuses it onto a detector. The detector produces an image covering 185 by 180 km, which is about the size of New York City and suburbs.

The ETM+ uses eight different filters. See Fig. 3.1. Two of these transmit in the visible region, and the others transmit in the infrared region. Most of the filters have a resolution of 30 m/pixel.

How can *Landsat 7* image nearly the entire Earth? It orbits Earth once every 99 min, and as a result, the ground trace shifts ~25° of longitude. After 1 day, it is ~13° of longitude from where it started. It continues moving around Earth for the second through sixteenth days, always missing its start position by a little. After 16 full days, this satellite flies over its initial path and the cycle starts over. The shift in

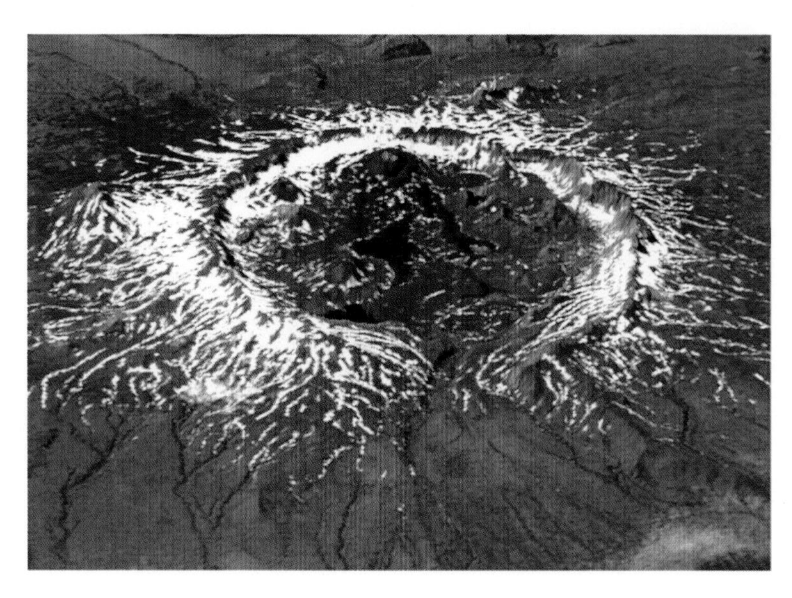

Fig. 3.2. An image of the Okmok volcano on Umnak Island, one of Alaska's Aleutian Islands. This volcano is at 54.4°N, 168.1°W. Scientists used *Landsat 7* and *Airsar* data to construct this image. This image is 16 km or 10 miles wide (Credit: NASA/JPL/ASF).

ground trace causes the satellite to follow 233 different paths. Furthermore, there are 248 places where it may record an image on each path. As a result, there are $248 \times 233 = 57{,}784$ different positions. Over the course of the 16-day cycle, it will have flown over most of the world. Two adjacent images overlap. This overlap insures that no area is missed. Since *Landsat 7* has an orbital inclination of 98.2°, it does not cross over the North and South poles. As a result, it does not image areas north of 83°N or south of 83°S. Figure 3.2 shows an image constructed from *Landsat 7* and *Airsar* data.

Although *Landsat 7* flies over most of Earth in 16 days, it does not image every one of the 57,784 different positions during each 16-day cycle. Instead, areas are prioritized. Several factors determine whether an area will be imaged. Cloud coverage plays a big role. An image showing just clouds will have little or no useful data. Therefore, when clouds are predicted to cover an area it is not imaged. For similar reasons, areas that have low Sun altitudes, such as those in northern Canada during the winter, receive a low priority. Areas that are not undergoing significant change, like those experiencing winter, are also given a low priority. On the other hand, areas with a large fire, a flood, or an active volcano receive a high priority. Similarly those that were missed in previous 16-day cycles receive a high priority. Areas experiencing rapid change such as what happens during the spring also receive a high priority.

With different filters, one may produce low resolution spectra from Landsat data. Using a combination of images and spectra, scientists may monitor the health of crops, estimate the effects of urban sprawl on the environment, and monitor water resources. Furthermore, since Landsat satellites have operated since 1972, scientists may study environmental changes covering several decades.

Landsat 7 also measures temperatures. It is able to do this because Earth emits infrared radiation with wavelengths between 8 and 12 μm. This overlaps with filter #6. Temperatures are measured in a similar way as is described in Chap. 1.

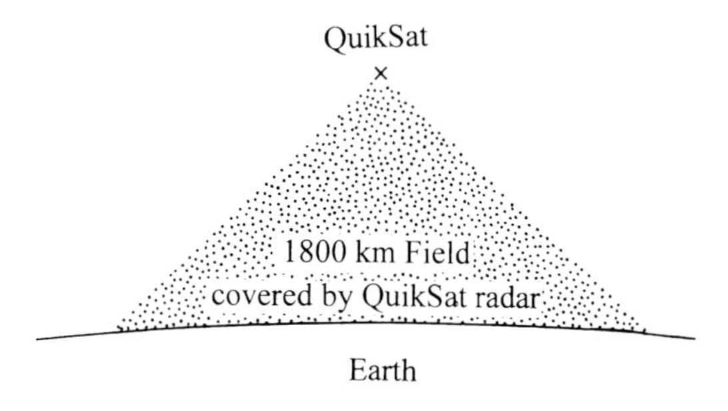

Fig. 3.3. The *QuikScat* satellite detects microwaves over an area 1,800 km wide (Credit: Richard Schmude, Jr.).

QuikScat

Meteorologists need wind data to forecast weather. Winds determine the movements of clouds and hurricane strength. They can also lead to wind shear, which is dangerous for aircraft. Therefore, wind data is of great importance. It is difficult to measure high-altitude winds and those over the ocean. This is where data from the *QuikScat* satellite becomes important. It emits pulses of radio waves with a frequency of 13.4 GHz. These hit Earth and bounce back. See Fig. 3.3. The backscattered radiation varies with the wind speed and direction. Therefore, scientists can examine the backscattered radiation and compute the wind speed and direction. They are able to measure winds in an area as small as 25 km across. Uncertainties in the wind speed and wind direction are about 12% and 20°, respectively.

Since *QuikScat* spins 18 times a minute, radar from several different angles strikes several locations. As a result, scientists are able to measure the wind speed and direction for almost any area of interest. It uses radio waves to collect wind data and can operate under almost all weather conditions.

In addition to yielding data on winds, *QuikScat* has many other applications. It can track icebergs. Most of an iceberg lies underwater. Furthermore, icebergs move with ocean currents. For these reasons, icebergs are a hazard to shipping. *QuikScat* data yields up-to-date coordinates of hazardous icebergs. It also yields information on the amount of moisture in the soil. See Fig. 3.4. This is important for the control of wildfires. Scientists may also monitor vegetation growth and changes.

Terra

Have you ever felt relief from the summer heat when a cloud passed in front of the Sun? Have you ever felt a cool sea breeze? Has an annoying vapor trail from a plane ever blocked your view of a star or planet? If you answered yes to any of these questions, you would have witnessed the interaction between the ocean, atmosphere and land. The *Terra* satellite has five instruments, which are designed to monitor the atmosphere, land masses, ice caps and bodies of water. More importantly, it gives scientists information on how the different parts of Earth's biosphere interact with one another.

Fig. 3.4. This image shows the soil moisture change between May 2 and May 3, 2009. *Green* shows moisture whereas *gray* shows a lack of moisture. The *QuikScat* satellite took both images. Soil moisture data may help firefighters learn more about wildfires and how to better contain them (Credit: NASA/JPL/QuikScat Science Team).

One specific application of *Terra* data is to determine how aerosols affect the atmosphere and the weather. Aerosols are microscopic particles that float in the atmosphere. Figure 3.5 shows an image of ship tracks. These form when a ship gives off aerosols that act as condensation nuclei for water vapor. Water vapor condenses on them, forming droplets. We see millions of these as long, thin clouds. The resulting cloud affects temperatures and winds. A similar process also occurs behind planes and in areas with large amounts of aerosols. One important question is: "Do aerosols affect the frequency of clouds?" Data from the *Terra* satellite should help answer this question.

A second application of the Terra satellite is to monitor the movement and temperature of large ice sheets. In some cases, large chunks of ice break off, becoming icebergs. Figure 3.6 shows a sequence of three images taken in November of 2000 that show a large chunk of ice breaking off from west Antarctica. Ice sheets have broken apart throughout Earth's history. The question: "Is there a change in the frequency of ice sheets breaking apart over time?" Scientists hope to answer this question with *Terra*.

The *Terra* satellite is able to peer down a dangerous volcano at a safe distance. *Terra* is able to measure temperatures. The high temperature of hot magma will

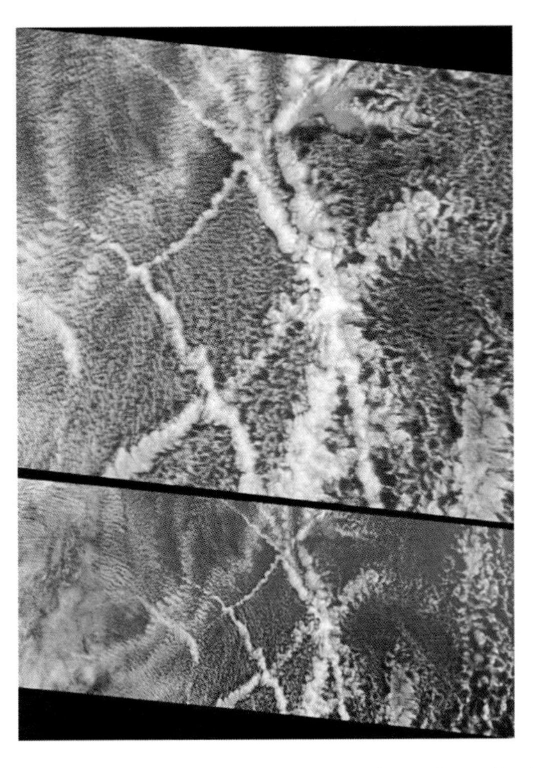

Fig. 3.5. This image shows ship tracks. These are clouds produced from fine particles (or aerosols) coming from ships. These particles serve as condensation nuclei, leading to the formation of clouds. The *Terra* satellite will help scientists learn more about how aerosols impact the weather (Credit: NASA/GSFC/LaRC/JPL, MISR Team).

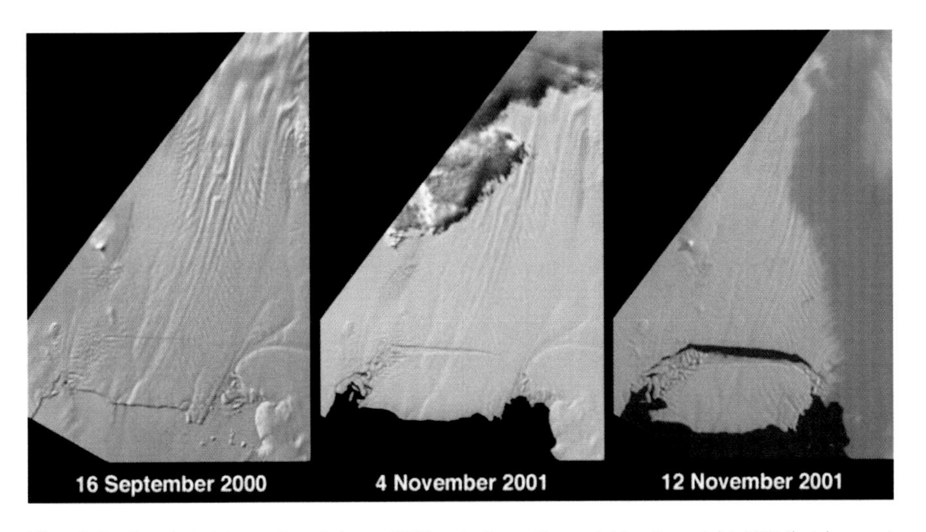

Fig. 3.6. The Multi-Angle Imagine SpectroRadiometer (MISR) on the *Terra* satellite recorded these images in late 2000. The iceberg was at 75°S, 102°W at the time it broke off. It is 42 km wide and 17 km from top to bottom. Between November 4 and 12, it moved at an average speed of 26 m per hour. One important application of Earth-observing satellites is to keep track of large icebergs (Credit: NASA/GSFC/LaRC/JPL, MISR Team).

Fig. 3.7. An image of the Russian volcano Shiveluch. The Advanced Spaceborne Thermal Emission and Reflection Radiometer (ASTER) on the *Terra* satellite recorded this image on January 25, 2011. This image was taken at night. It is 6.7 by 6.7 km across. The *bright spot* just above the center is the center of the volcano. It is bright because it is hot (Credit: NASA/GSFC/METI/ERSDAC/JAROS and U. S./Japan Aster Science Team).

stand out in a *Terra* image, as is shown in Fig. 3.7. This image shows the Russian volcano Shiveluch on January 25, 2011. The bright spot just above the center is hot magma. This satellite will enable scientists to pinpoint the path of any approaching molten rock.

Terra is one of several satellites that are in the same orbit and follow one another. These are called the A-Train. At one point, the A-Train had six satellites: *Aqua* (leading satellite), *Terra, Aura, CloudSat, Calipso* and *Parsol.* Each of these carried out a different measurement. This allowed scientists to better understand processes which are at work in Earth's oceans, atmosphere and land masses.

ACRIMSAT

In the last few years there has been a lot of debate about climate change, global heating and greenhouse gases. If global heating is real, what is causing it? Is it caused solely by the increase in greenhouse gases in our atmosphere or are other factors responsible? Many believe this is a complex issue that needs to be examined thoroughly before laws are enacted. One factor which undoubtedly drives climate change is the variability of our Sun. Table 3.2 lists a few events that affect the amount of sunlight hitting Earth. The *Active Cavity Radiometer Irradiance Monitor Satellite (ACRIMSAT)* is designed to measure the amount of energy coming from our Sun.

The *ACRIMSAT* satellite measures the total amount of energy coming from the Sun; this is called Total Solar Irradiance, or TSI. It takes measurements with an accuracy of 0.01%. We know the TSI (or solar constant) is ~1,366 W/m^2 at Earth. This is the amount of solar energy striking 1 m^2 of the top of our atmosphere when

Table 3.2. Events that affect the amount of solar energy striking the top of our atmosphere

Event	Percentage change	Source
11-year sunspot cycle	0.1–0.2	Figure 3.8
Eclipse of Moon	Up to 99.5	James and Mason (2003)
Transit of Venus	~0.1	Schneider et al. (2006)
Transit of Mercury	~0.004	Author's calculation
Flare	?	–

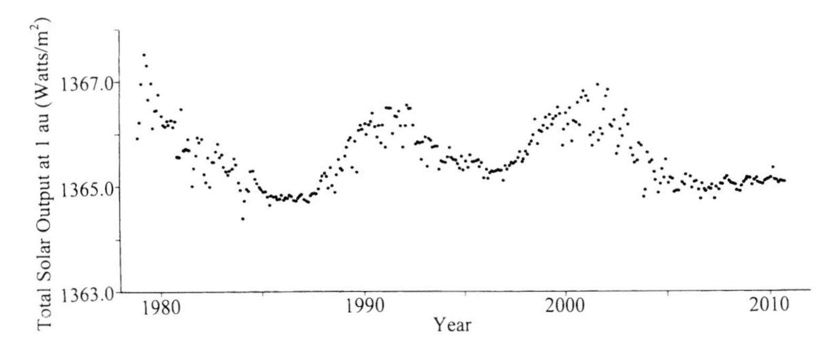

Fig. 3.8. The total solar output between 1978 and 2010 for a distance of 1.00 astronomical unit is plotted here. Each point is the average output for 0.10 years. The average deviation, d, of each point is about 0.07 W/m^2. In equation form, $d = s/(N)^{1/2}$ where σ is the standard deviation and N is the number of measurements made over the 0.1 year period. The value of N was usually between 34 and 37 (Credit: NASA, The ACRIMSAT team for providing the data on their website, http://lasp.colorado.edu/source/data/tsi_data.htm#summary_table, and to Richard Schmude, Jr., who computed the average values and constructed the figure).

the Sun is directly overhead. This number, however, changes. Since 1978, scientists have used satellites to measure the TSI. A compilation of these measurements is listed on the *ACRIMSAT* website. Measurements were usually made about once a day. The average values of the TSI for every 0.1 year is compared. Figure 3.8 shows the results. Between 1978 and 2010, the TSI changed. It reached maximum values in 1979, 1991 and 2000, and it reached minimum values in 1986, 1996 and 2007. The difference between the maximum and minimum values was up to 3.0 W/m^2. This is ~0.2% of the TSI. Willson and Mordvinov (2003) report an increase of 0.05% per decade. This undoubtedly affects our climate. Future *ACRIMSAT* data should enable astronomers to learn more about how our Sun drives climate change.

Earth Observing-1 (EO-1)

The three instruments on *EO-1* are the Advanced Land Imager (ALI), the Hyperion High Resolution Spectral Imager (Hyperion) and the Linear Etalon Imaging Spectrometer Array atmospheric collector (LAC). The ALI has ten different filters that span wavelengths of between 0.48 and 2.35 μm. It is capable of yielding images with a resolution of 10–30 m depending on the filter used. At an altitude of 705 km, it images a 20 by 180 km area. The Hyperion instrument is able to construct images in 220 different spectral bands. Scientists are also able to collect spectra with it. Images are 7.5 by 100 km and have resolutions of 30 m. Figure 3.9 shows three

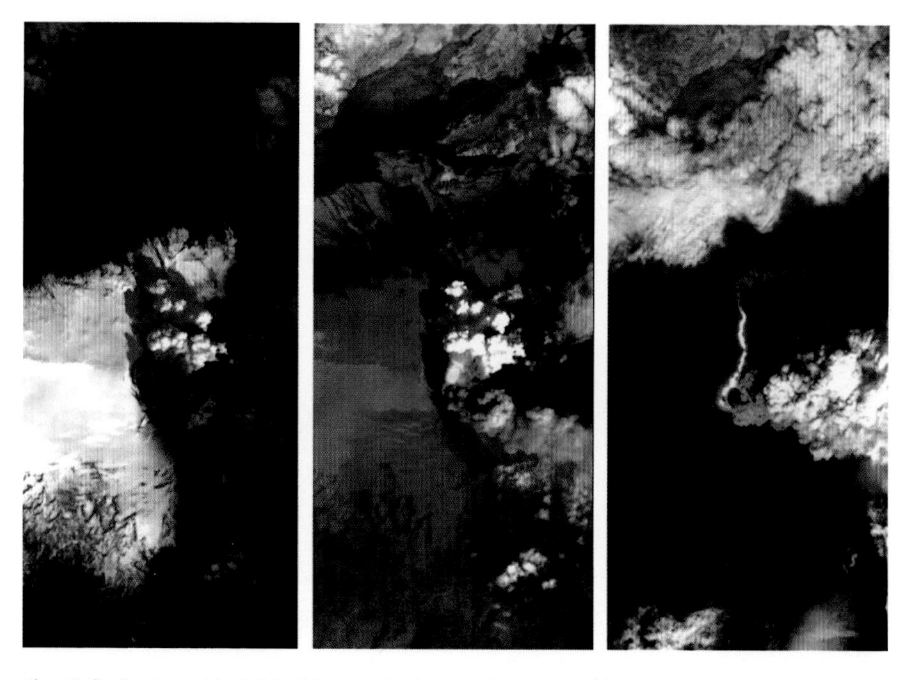

Fig. 3.9. Three images of the Eyjafjallajökull icecap and nearby volcano. The image on the *left* is a true color image showing *white* ice and the *gray* ash cloud. The *middle image* shows the same volcano in near-infrared light. The *blue* areas are ice and the *red* and *yellow* area is a hot spot. The image on the *right* shows the lava flow (*yellow*) on May 4, 2010. The Earth Observing *(EO-1)* satellite took all three images (Credit: NASA/JPL/EO-1 Mission/GSFC Ashley Davis).

images of the Eyjafjallajökull volcano and ice cap in Iceland. These images enable scientists to learn about its ash and steam clouds. The LAC instrument collects data of Earth's atmosphere. This is used to calibrate measurements taken with the other instruments.

During the first year of operation, EO-1 was a technology demonstration mission. Essentially, the ALI was lighter and used less power than older cameras. It performed well enough that an extended mission was funded. As of early 2011, this satellite is still operating. Some of the applications of it include imaging areas affected by natural disasters, mineral mapping and estimating crop yield.

Jason-1

Any increase in sea level may make coastal areas inhabitable. For this reason, it should be watched. A rise in temperature or the melting of ice on land will cause ocean levels to rise. *Jason-1* and its successor, *Jason-2,* are designed to measure sea level to an accuracy of ~2.5 cm. These satellites take measurements over most of the Atlantic, Indian and Pacific oceans.

The *Jason-1* satellite has four instruments that measure its exact position and velocity. This is needed to record accurate measurements of sea level. These four are the Doppler Orbitography and Radiopositoning Integrated by Satellite (Doris), the Jason Microwave Radiometer (JMR), the Turbo Rogue Space Receiver (TRSR) and the Laser Retroreflector Array (LRA). The Doris instrument is used to measure the

Fig. 3.10. This false color image shows different altitudes of the ocean. *Blue* indicates the ocean is about 10 cm lower than normal. This is consistent with lower temperatures that are attributed to La Niña. The *Jason-2* satellite recorded this data on June 11, 2010. Images like this one will allow meteorologists to make weather forecasts several weeks into the future (Credit: NASA/JPL Ocean Surface Topography Team).

exact velocity of *Jason-1* using the Doppler effect. This is described in Chap. 1. The JMR measures the amount of water vapor in the troposphere. This affects the computed satellite altitude by up to 50 cm. The TRST instrument has a receiver that picks up Global Positioning Satellite signals from navigation satellites. These enable scientists to determine *Jason-1's* orbit to a high degree of accuracy. A fourth instrument, the LRA, is a system of mirrors. Laser pulses from the ground are directed at the LRA, which reflects the light back. By measuring the length of time for the laser light to return, scientists measure the altitude of *Jason-1* to within a few millimeters. This serves as a calibration for the Doris, JMR and TRSR measurements.

The fifth instrument, the Poseidon-2 altimeter, measures the distance between the satellite and the ocean surface. It sends out pulses of radio waves that bounce off the ocean and come back. Scientists measure the time it takes the pulses to come back and compute the distance. With accurate satellite position, scientists may determine the ocean levels to a high degree of accuracy.

Two follow-up missions to *Jason-1* are *Jason-2* and the *Ocean Surface Topography Mission (OSTM)*. Both satellites will monitor the level of the oceans over several years.

What have scientists found out from *Jason-1* and *Jason-2* data? We know the oceans are not exactly level at large scales even after the effects of lunar and solar tides are eliminated. The affects of waves on scales of a few feet are averaged out over many miles. Figure 3.10 shows a *Jason-2* false-color image of the Pacific Ocean. The colors correspond to altitudes. Purple and blue are low altitudes. Red and orange are high altitudes. Apparently, the Pacific Ocean level changes by several centimeters. The bluish areas are about 5–13 cm below the average level. This is caused by lower-than-average ocean temperatures. Those low temperatures lead to the La Niña effect.

Aqua

Aqua carries out several types of atmospheric measurements. These are made at different altitudes. It also measures ocean temperatures.

Aqua contains six different instruments. One of them, the Advanced Microwave Scanning Radiometer for the Earth Observing System or AMSR-E, measures the microwaves coming from Earth's surface. Once correction factors are applied, scientists may measure the quantity of microwaves being emitted. This yields land and ocean temperatures. Since this instrument is based on microwaves, it operates under cloudy and clear conditions. The only time that it cannot collect data is when lots of rain or snow are falling. The conditions under which the AMSR-E can (and cannot operate) are illustrated in Fig. 3.11.

A second instrument on *Aqua* is the Atmospheric Infrared Sounder (AIRS). It measures spectra between wavelengths of 0.4 and 15.4 μm. These yield temperatures at different altitudes in our atmosphere. These are helpful for weather forecasting.

One specific application of AIRS is to measure the temperature of hurricanes and typhoons. Figure 3.12 shows a false color image of the typhoon Yasi as it passed over the east coast of Australia. Colors correspond to different temperatures. Red and orange correspond to high temperatures. Purple and blue correspond to lower ones. This image also shows Yasi still retained its spiral structure after it made landfall.

The *Aqua* satellite has four other instruments, namely the Advanced Microwave Sounding Unit-A (AMSU-A), Cloud and Earth's Radiant Energy System (CERES), Humidity Sounder for Brazil (HSB) and the Moderate Resolution Imaging Spectroradiometer (MODIS). These are designed to detect water vapor in the atmosphere and carry out cloud measurements. This data helps scientists learn more about Earth's water cycle.

This satellite will also keep track of Earth's radiation budget. Let's explore what this is.

The radiation budget summarizes the amount of energy reaching and leaving Earth. If the amount given off is less than what is coming in, Earth will get hotter; otherwise it will get colder. Sunlight, mostly in the form of visible light, reaches half of Earth at any given time. The light reaching Earth warms up the atmosphere. Some of it also warms up the surface. Light leaves Earth through three processes: refraction/scattering, reflection and emission. Our atmosphere refracts and scatters about 1% of the sunlight. This is discussed further in Chap. 6. About one-third of it is reflected. The remaining 60–70% is emitted mostly as infrared radiation. The radiation budget is illustrated in Fig. 3.13.

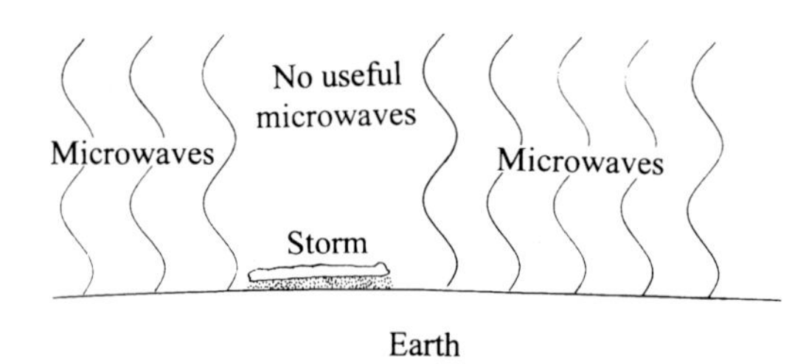

Fig. 3.11. The Advanced Microwave Scanning Radiometer for the Earth Observing System (AMSR-E) on the *Aqua* satellite measures the number of microwaves that Earth's land and oceans emit. Rain and snow prevent useful microwaves from reaching the satellite (Credit: Richard Schmude, Jr.).

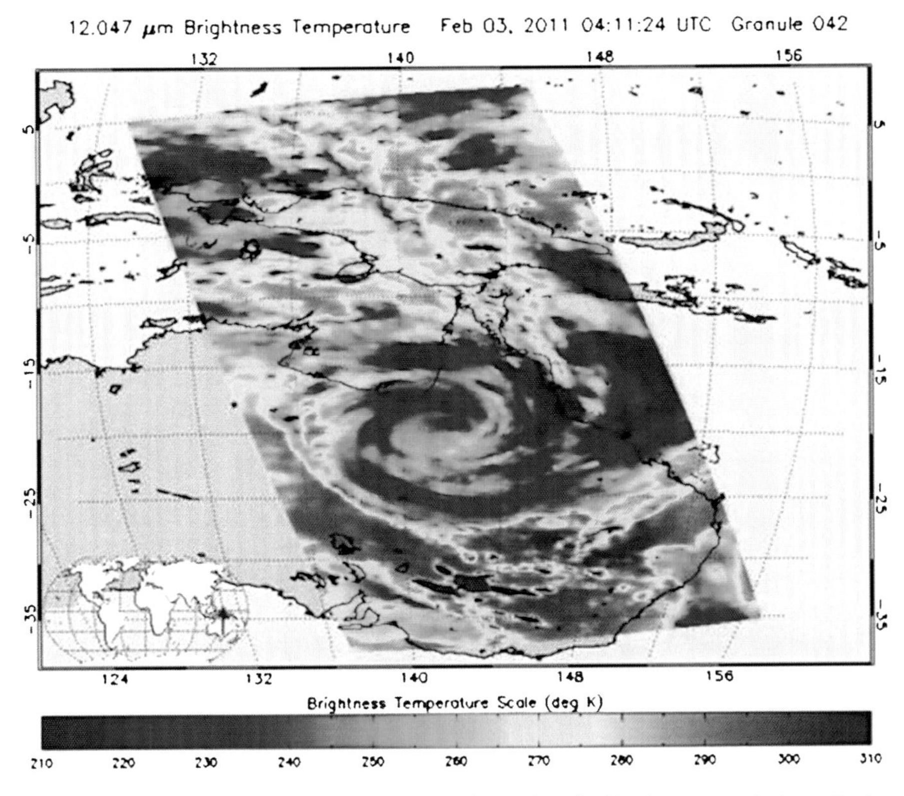

Fig. 3.12. The temperatures inside the cyclone Yasi measured with the Atmospheric Infrared Sounder instrument on the *Aqua* satellite. *Aqua* recorded this image on February 3, 2011. Images like this one will enable meteorologists to learn more about cyclones (Credit: NASA/JPL-Caltech).

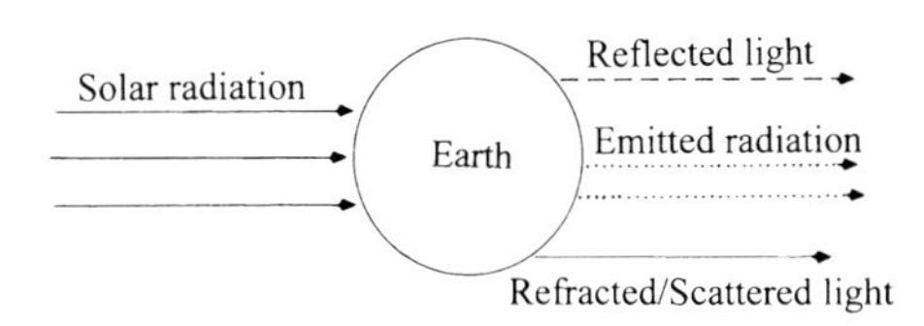

Fig. 3.13. This is a simplified version of Earth's radiation budget. Clouds and the surface reflect just over a third of the solar radiation. Most of the remainder is absorbed and later emitted as infrared radiation. Our atmosphere refracts and scatters about 1% (Credit: Richard Schmude, Jr.).

There are several factors that may affect Earth's radiation budget. For example, the percentage of reflected sunlight depends on the amount of cloud cover along with the amount of snow and vegetation present. Other factors such as clouds, snow and ice cover, the relative humidity and changes caused by human activity also affect Earth's radiation budget. The amount of greenhouse gases may also affect it. In fact, scientists are interested in knowing whether the amount of infrared radiation emitted into space is changing with the rising carbon dioxide levels. Some scientists believe human-made changes to our atmosphere are causing the amount of escaping heat to drop, leading to global heating.

Data from the various instruments on *Aqua* will enable scientists to learn how the radiation budget affects our weather and climate. It will also help us learn whether this budget is changing.

GRACE

In physics class, you probably learned that Earth's gravitational acceleration at sea level equals 9.81 m/s^2. This is correct; however, Earth's gravity changes a little. The gravity changes by 0.001–0.01% over a period of hours to years. Some of the factors that cause it to change include rain, snowfall, the drying up of the ground, earthquakes, the growth or shrinkage of aquifers, the growth or loss of vegetation and the growth of lakes next to a dam. A redistribution of water vapor in our atmosphere may also cause a change in gravity. In short, any redistribution of mass will cause a change in the local gravity. The *GRACE* satellite system is designed to measure Earth's gravity at different locations. We will now learn how scientists make these measurements.

The distance between the two identical *GRACE* satellites yields gravity data. They follow the same orbit and are about 220 km apart. Each contains GPS equipment along with microwave ranging instruments. The equipment allows scientists to compute the exact distance between them. Figure 3.14 illustrates how scientists detect a gravity change. In the top frame, the satellites are distance X apart. In the bottom frame, the leading satellite approaches an area with a higher than normal gravity. The leading satellite accelerates more to this area than the trailing one. This is because the gravitational field increases with the inverse square of the distance. As a result of the difference in acceleration, the distance between them is now $X + \Delta x$. Scientists use the Δx value to compute the difference in gravity.

One application of *GRACE* data is to monitor the seasonal rainfall in specific areas. For example, Tapley et al. (2004) report the Amazon Basin had a higher gravity in April and May of 2003 than what it had been 3 months earlier. This is undoubtedly caused by the large amount of rainwater that accumulated. See Fig. 3.15. The red and reddish pink areas have higher than normal gravity because of excess water.

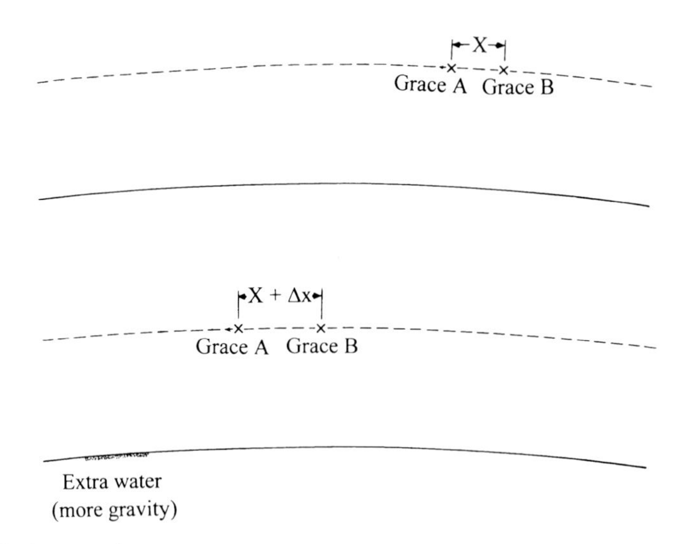

Fig. 3.14. The GRACE satellites are initially × km apart. As the leading GRACE satellite *A* approaches an area with extra water, it accelerates more than the trailing satellite *B*. This causes the satellites to be $X + \Delta x$ km apart. Scientists use the change in distance to determine the extra gravitational acceleration and, hence, the mass change (Credit: Richard Schmude, Jr.).

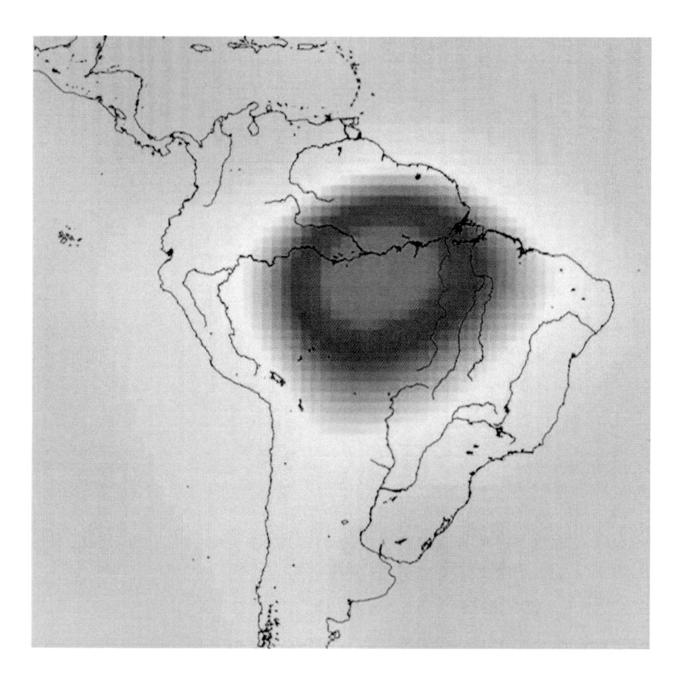

Fig. 3.15. A false color image showing the amount of water in the Amazon basin in South America. The *pink* and *red area* shows an excess of water. The GRACE satellite measures tiny changes in Earth's gravity caused by water near the surface. This image was made during the northern spring season of 2008. This data may help scientists learn more about groundwater and aquifers (Credit: NASA/JPL/University of Texas Center for Space Research).

Scientists have used *GRACE* data to monitor aquifers. Aquifers are large bodies of water that lie underground. Rodell et al. (2009) report that *GRACE* detected a drop in Earth's gravity in northern India. After ruling out other factors such as irregularities in average annual rainfall, they report groundwater depletion is the best explanation for the gravity drop. The *GRACE* data are consistent with a loss of 109 cubic km of groundwater between 2002 and 2008. See Fig. 3.16.

Scientists also use *GRACE* data to monitor the thickness of large ice sheets. Figure 3.17 shows the changes in Greenland's ice cap thickness between 2003 and 2008. As the cap gets thinner, the gravity drops. In many areas, it is getting thinner. These results are based on gravity measurements.

ICESat

ICESat is designed to measure the altitude of ice both at sea and on land, provide altitude measurements of clouds and yield data on vegetation. It has Global Positioning System equipment along with other instruments that enable scientists to determine its exact position. It also has a 1-m telescope along with lasers that emit light with wavelengths of 532 and 1,064 nm. The 1,064 nm light is for sea and land ice, whereas the 532 nm light is for atmospheric features. When sea ice is the target, the satellite emits pulses of 1,064 nm laser light. It strikes the ice and the signal bounces back. The telescope detects the return signal. With exact return times, scientists can compute the distance between *ICESat* and the ice. By knowing

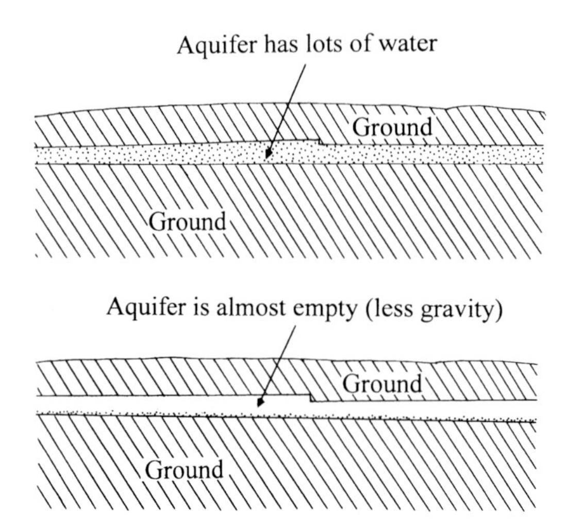

Fig. 3.16. The aquifer in the *top frame* has lots of water. As water is removed from the aquifer an underground void develops, as is shown in the *bottom frame*. This leads to a drop in gravity. The GRACE satellites are able to detect this (Credit: Richard Schmude, Jr.).

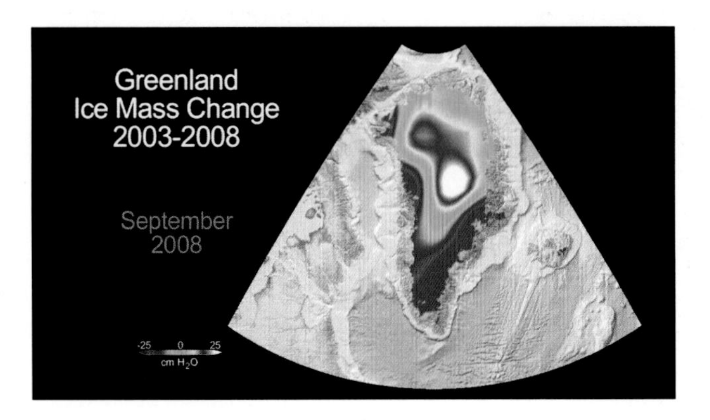

Fig. 3.17. Changes in the icecap thickness across Greenland between 2003 and 2008. These changes are based on gravity measurements made by the GRACE satellites. The *pink* are areas where the ice has gotten a few inches thinner. The *orange* and *white* areas are where the ice has gotten a few inches thicker (Credit: NASA/JPL).

the exact satellite position, they may determine the altitude of the ice surface to an accuracy of 10 cm. Repeated measurements improve this accuracy.

Two events that limit *ICESat* are thin clouds and snowfall. For example, Yang et al. (2010) point out that thin clouds may affect the 1,064 nm laser data by several centimeters. A second factor that may affect *ICESat* data is snowfall. This is because the laser light bounces off the top of the ice/snow layer. If a lower-than-average amount of snow has fallen, the data would show the altitude of the ice dropping even when the ice altitude is the same. See Fig. 3.18.

Farrell et al. (2009) carried out a study of sea ice in the Arctic Ocean between 2003 and 2008. These three used ICESat laser altimetry data to measure the altitude of sea ice. They report the altitude dropped by 1–2 cm per year. These scientists believe snowfall may have affected the results.

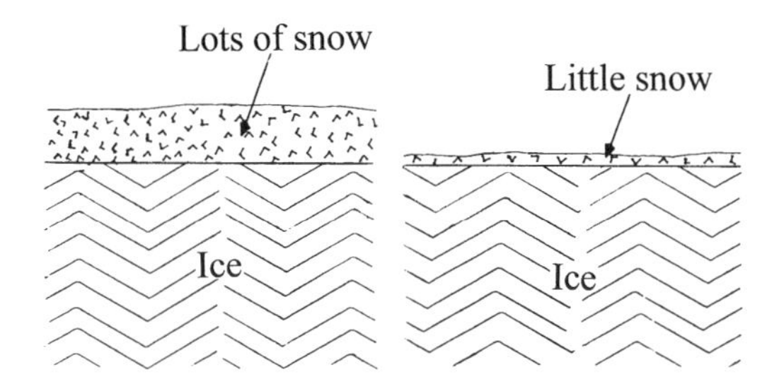

Fig. 3.18. The amount of snowfall may affect the *ICESat* results. A heavy snowfall means a higher snow level, as is shown on the *left*. A low snowfall causes a lower level, as is shown on the *right*. The ice level is the same in both frames. Therefore, a low snowfall may lead to a greater distance between the satellite and the ice. This is because *ICESat* only measures the distance to the top of the snow level (Credit: Richard Schmude, Jr.).

SORCE

The Solar Radiation and Climate Experiment (SORCE) satellite monitors the total amount of radiation coming from the Sun at specific wavelengths. It has four instruments. The TIM instrument measures the total amount of solar energy. It has an accuracy of 0.01%, and after repeated measurements, the accuracy improves. It detects small changes in solar output. TIM instruments on previous satellites collected the data plotted in Fig. 3.8. The other three instruments on the *SORCE* satellite measure the amount of solar energy being emitted at specific wavelengths. Scientists are interested in learning how the different wavelengths of sunlight affect our climate.

Rottman et al. (2005) report preliminary *SORCE* data taken in July and August of 2004. During this time, the total solar output changed by about 0.1%. The change in ultraviolet light with a wavelength of 280 nm was even higher at 0.3% in July 2004.

Aura

The *Aura* satellite detects trace quantities of ozone, methane, water vapor, hydrogen cyanide (HCN) and other gases in our atmosphere. Its four instruments are the High-Resolution Dynamics Limb Sounder (HIRDLS); the Microwave Limb Sounder (MLS); the Ozone Monitoring Instrument (OMI) and the Tropospheric Emission Spectrometer (TES). These will help answer several questions such as: (1) Is the amount of stratospheric ozone continuing to decline? (2) What role does the Sun's ultraviolet emission play in stratospheric ozone levels? (3) Are the amounts of greenhouse gases, such as water vapor and methane, continuing to rise? A greenhouse gas is one that is transparent to visible light but is opaque to infrared radiation with wavelengths of around 10 μm. Essentially, these gases trap heat much like a greenhouse.

Before we discuss the *Aura* satellite, we will present some facts about ozone.

Ozone is a form of oxygen. The oxygen we breathe contains two oxygen atoms and is written as O_2. A molecule of ozone contains three oxygen atoms and is written as O_3. We need molecular oxygen to survive, but ozone is a poison. However, ozone is also beneficial because it absorbs harmful ultraviolet radiation. Two ozone layers exist. One lies near the surface of Earth and the other lies in the stratosphere.

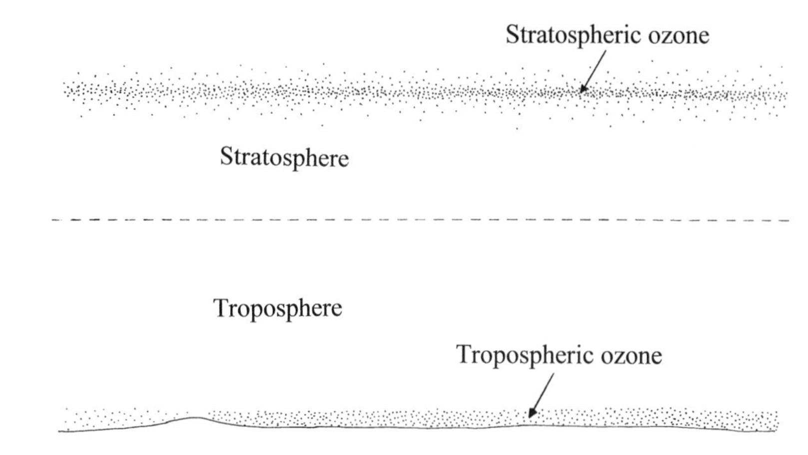

Fig. 3.19. The tropospheric ozone layer lies in the troposphere and the stratospheric ozone layer lies in the stratosphere. The *dashed line* is the boundary between the troposphere and the stratosphere (Credit: Richard Schmude, Jr.).

See Fig. 3.19. The layer near the ground is called tropospheric ozone and the other one is called stratospheric ozone. Tropospheric ozone is harmful, and it arises from a combination vehicle traffic and the burning of coal and other combustible materials. Stratospheric ozone, however, is beneficial because it absorbs harmful ultraviolet radiation.

The biggest danger to humans is the destruction of the stratospheric ozone layer. Without it, more destructive ultraviolet radiation will reach the surface. During the 1980s, this layer was deteriorating rapidly. Most scientists believe this was caused by the emission of chlorofluorocarbons. During this time, limits were placed on these compounds. One of the goals of the *Aura* satellite is to monitor the amount of stratospheric ozone.

The amount of solar ultraviolet radiation also affects this layer. Haigh et al. (2010) report a drop in solar ultraviolet radiation may have caused a drop in stratospheric ozone below 45 km but caused an increase in this gas at higher altitudes. This group used data from the *SORCE* satellite to monitor solar ultraviolet emission and used *Aura* data to reach this conclusion. In addition to ozone, the *Aura* satellite monitored an isotope of water.

Figure 3.20 illustrates the amount of HDO measured by *Aura* on January 6, 2006. HDO is an isotope of water. It has a deuterium atom in place of hydrogen. The depletion is greatest near Labrador and northeastern Quebec, Canada, and at altitudes above 6 km. This compound is used to trace the history of water vapor in our atmosphere. As it turns out, water is about 0.03% HDO. When it rains, HDO is more likely to fall than H_2O. As a result, the amount of HDO in the atmosphere will be lower than 0.03% after a rainstorm. Therefore, the drop near Labrador and northeastern Quebec in this illustration is probably caused by large amounts of precipitation.

Calipso

The Cloud, Aerosol, Lidar and Infrared Pathfinder Satellite Observation (Calipso) is a French-American satellite designed to study clouds and aerosols. It will enable

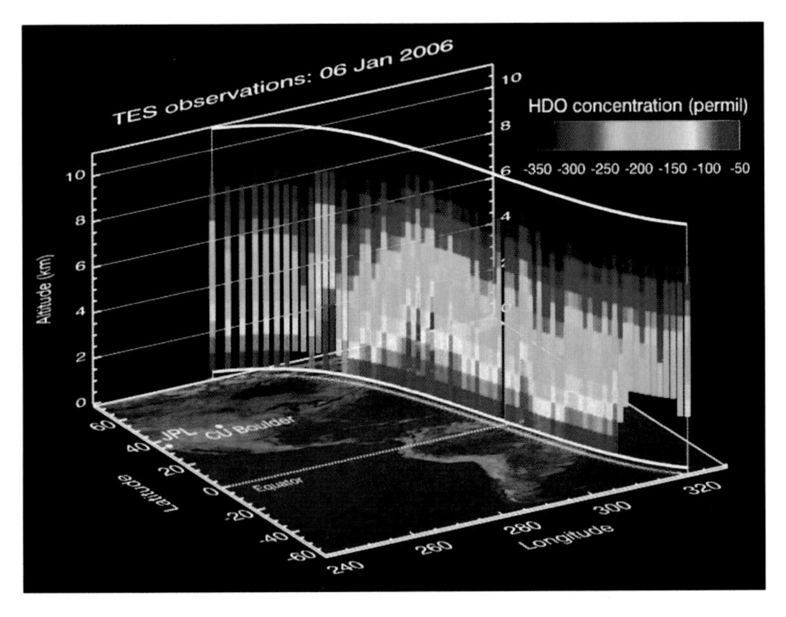

Fig. 3.20. A graph illustrating the concentration of the water isotope HDO across the path *Aura* followed. The *Aura* satellite recorded this data on January 6, 2006. The HDO concentrations will help scientists learn more about the water cycle and its effect on weather and climate. Note that the longitude is measured eastward from the Prime Meridian. Therefore, 240° longitude in this diagram equals 120°W longitude (Credit: NASA/JPL-Caltech).

scientists to take measurements on two or more layers of clouds and to determine how thick aerosol layers are. Scientists are especially interested in learning how clouds and aerosols affect the temperature. *Calipso* contains a 1-m telescope along with visible and infrared imagers and a lidar. The lidar illuminates clouds and aerosols with polarized laser light and then detects the reflected and scattered light. When the signal interacts with the atmosphere, the polarization changes, and it is this change that may yield information about the particles making up clouds and aerosols. The laser has wavelengths of 532 and 1,064 nm.

Cloudsat

Clouds reflect sunlight. They cause less sunlight to reach the surface and, hence, lead to cooler daytime temperatures. At night, clouds prevent heat from escaping, causing temperatures to remain high. To make matters more complicated, the thickness of clouds affects how much light makes it to the surface or how much heat is trapped. In short, clouds affect our temperature. The *Cloudsat* satellite is designed to study the frequency, distribution, structure and radiative properties of clouds. It contains a 94 GHz radar instrument that directs radar downwards. The radar bounces off clouds, and scientists use the return signal to learn more about their distribution and thickness. In many cases, data from this satellite is used with those from others to determine how clouds affect local conditions. The vertical resolution of the radar instrument is 0.5 km or 0.3 miles. The spatial resolution is 1.4 by 1.7 km, which is about a square mile.

Other Earth-Observing Satellites

Many more Earth-observing satellites are in operation than those listed in Table 3.1. Many of these have specific functions. A group of satellites designed to study the Earth's land masses is summarized in Table 3.3. Tables 3.4 and 3.5 list several Earth observing satellites operated by the European Space Agency and their partners.

Weather Satellites

The weather is the state of the atmosphere at a specific time and place. This includes the temperature, pressure, relative humidity, winds, cloud cover and so forth. Climate is different. It is the average weather condition for many years at a location (Gove, 1971). Satellites whose primary purpose is to collect data on our weather are called weather satellites. They provide nearly continuous data on the state of our atmosphere. Several countries operate them. See Tables 3.6 and 3.7.

In recent years, our dependence on satellites, large power grids and pipelines means that forecasts of space weather are also vital. Space weather includes the

Table 3.3. Selected satellites that image the surface of Earth

Satellite	Year launched	Orbit			Minimum ground track repeat cycle (days)	Source
		Type[a]	Altitude (km)	Inclination (degrees)		
Landsat 7	1999	LELE-SS	705	98.2	16	b
ASTER	1999	LELE-SS	705	98.3	16	c
Ikonos	1999	LELE-SS	681	98.1	3	c
QuickBird	2001	LELE-SS	450	97.2	1	c
SPOT-5	2002	LELE-SS	822	98.7	2–3	c
CBERS-2	2003	LELE-SS	778	–	5	c, d
FORMOSAT-2	2004	LELE-SS	894	99.1	1	c, e, f
CARTOSAT-1	2005	LELE-SS	618	Polar	116	c
ALOS	2006	LELE-SS	~700	98	46	c, g
RADARSAT-2	2007	LELE	798	98.6	24	h
WorldView-1	2007	LELE-SS	496	–	1.7	c
GeoEye-1	2008	LELE-SS	684	98	2.1	c
WorldView-2	2009	LELE-SS	770	–	1.1	

[a] *LELE-SS* low-Earth low eccentricity-Sun-synchronous
[b] http://eospso.gsfc.nasa.gov (click on mission profile and then Eos)
[c] http://www.satimagingcorp.com/
[d] http://space.skyrocket.de/doc_sdat/cbers-1.htm
[e] http://space.skyrocket.de/doc_sdat/rocsat-2.htm
[f] http://www.spotimage.fr/web/en/977--formosat-2-images.php
[g] http://www.nasa.gov/ (keyword Alos)
[h] http://www.radarsat2.info/

Table 3.4. Earth -observing satellites operated by the European Space Agency

Satellite	Year launched	Objectives	Source
ERS-2[a]	1995	Monitor ozone levels, measure ocean and surface temperatures, measure winds at sea	b
Proba-1[c]	2001	Technology demonstration mission; monitor crops and forests	b
Envisat	2002	Image Earth's land, atmosphere, oceans and ice caps	b
MSG-1	2002	Monitor weather	b
MetOp	2006	Monitor weather	b
GOCE[d]	2009	Measure Earth's gravity at different locations	b
SMOS[e]	2009	Measure soil moisture; measure ocean salinity at different locations	b
CryoSat-2	2010	Measure thickness of sea ice, image continental ice sheets	b
SWARM	2012	These three satellites will measure the strength and direction of Earth's magnetic field	b
ADM-Aeolus	2013	Measure winds on a global scale	b
Earth CARE	2015	Characterize clouds and aerosols; determine how clouds and aerosols affect climate change	b

[a]ERS stands for European Remote Sensing satellite
[b]http://www.esa.int/esaEO/index.html (under ESA's Earth-observing missions)
[c]Proba stands for Project for OnBoard Autonomy
[d]GOCE stands for Gravity Field and Steady-State Ocean Explorer
[e]SMOS stands for soil moisture and ocean salinity

state of our magnetic field near the surface along with the charged particle and high-energy radiation near our satellites.

The National Oceanic and Atmospheric Administration (NOAA) in the United States oversees several weather satellites. These are listed in Table 3.7. The National Weather Service uses data from these to issue weather forecasts. Many of these include severe weather warnings. They also monitor space weather.

The *Geostationary Operational Environmental Satellites (GOES)* monitor weather in the United States and surrounding areas. They are in a geostationary orbit, and hence, move at the same rate as Earth. They remain above nearly the same location. More than one satellite is needed to monitor weather across the United States. Much of the information on this satellite is from Space Systems-Loral, Reference #S-415-19 (GOES I-M Data Book), which is on the website http://goes.gsfc.nasa.gov/text/goes.databook.html.

Figure 3.21 shows the dimensions of the *GOES* satellite. Its solar array has a total surface area of 12.7 m^2. It produces between 1,050 and 1,300 W of power depending on the season. The solar array charges up the nickel-cadmium batteries. A solar sail is at the other end. It balances the radiation pressure exerted on the solar array. This helps keep the satellite oriented properly. *GOES* is about as long as two school buses. The central box is around 2.4 m across, which is about the size of a large golf cart. It contains the imager, sounder, high-energy particle detectors, and an X-ray imager. A brief description of the scientific instruments follows.

The imager collects data on cloud altitude, sea surface temperature, the amount of water vapor in the atmosphere and cloud cover during both the day and night. These characteristics control our weather. The imager has a 0.311 m Cassegrain telescope. The imager uses five different filters that define the five channels. It records an image of the entire Earth or part of it. The advantage of imaging a

Table 3.5. Earth-observing satellites in partnership with the European Space Agency

Satellite (partner)[a]	Year launched	Objectives	Source
Landsat 4, 5, 7	Several	Image the land masses of Earth	b
SeaWiFS (GeoEye)	1997	Monitor color changes in the ocean; identify pollution and marine phytoplankton concentrations	b
SPOT-4 (CNES)	1998	Image selected areas of Earth; monitor crops and plants; monitor ecosystems	b, c
Ikonos-2 (USA)	1999	Collect multispectral images of Earth at resolutions of 1 and 4 m	b
KOMPSAT-1 (KARI)	1999	Image selected areas of Earth	b
QuikSCAT (USA)	1999	Measure surface winds over the oceans	b
Terra (USA)	1999	Provide global data on the atmosphere, land and oceans; monitor global climate change, collect data on how humankind impacts the environment; disaster prediction	b
Odin ([d])	2001	Collect data on stratospheric ozone and Earth's upper atmosphere	b
Aqua (USA)	2002	Measure changes in the chemical composition of the atmosphere; monitor stratospheric ozone levels	b
DMC (United Kingdom)	2002	Provide daily global imaging capability for rapid- response disaster monitoring and mitigation	b
IRS-P6 (ISRO)	2003	Provide images of land and water resources	b
SciSat-1 (CSA)	2003	Measures the abundances of trace gases in Earth's atmosphere	b
FORMOSAT-2 (Taiwan)	2004	Image specific areas on Earth	b
NOAA 18 (USA)	2005	Monitor the weather	b, e
ALOS (JAXA)	2006	Survey natural resources and natural disasters; image land masses	b
KOMPSAT-2 (KARI)	2006	Provide surveillance of large scale disasters survey natural resources	b
GOSAT (JAXA)	2009	Measure the amounts of carbon dioxide in different parts of the world; determine sources and sinks of this gas	b, e

[a]*CNES* Centre National d'Etudes Spatiales, *CSA* Canadian Space Agency, *ISRO* Indian Space Research Organization, *KARI* Korea Aerospace Research Institute

[b]http://www.esa.int/esaEO/index.html (go to ESA's Earth observing missions and then to Third Party Missions overview)

[c]http://spot4.cnes.fr/spot4_gb/index.htm

[d]Includes Canada, Finland, France and Sweden

[e]Use website in footnote a and then click on Third Party Mission technical site

small area is it takes less time. Therefore, meteorologists use it to focus on a rapidly developing storm. The imager plays an important role in weather forecasting; however, data from the sounder also provides vital data.

The sounder yields information on the vertical profile of our atmosphere. It measures how the temperature, ozone and water vapor levels change at different altitudes. This instrument also measures cloud temperatures. The sounder is able to do this through its 19 different channels. Each one uses a special narrow-band filter which is sensitive to either red or near-infrared wavelengths. This instrument is able to image the entire Earth or a small part of it.

Some factors that may affect space weather include solar flares, geomagnetic storms and bursts of high-energy particles slamming into our atmosphere. These could wreck havoc. For example, rapid changes in our magnetic field may create unwanted electrical currents in a power grid or a pipeline. This could lead to a

Table 3.6. Weather satellites operated by Europe and other countries besides the United States

Satellite	Country or group	Year launched	Orbit	Source
KALPANA-1	India	2002	Geostationary (74°E)	a
INSAT-3A	India	2003	Geostationary (93.5°E)	a
MTSAT-1R	Japan	2005	Geostationary (145°E)	b
MTSAT-2	Japan	2006	Geostationary (140°E)	b
METOp	Europe	2006	Inclination = 98.7°; altitude = 817 km	c
Fenyun-3	China	2008	Polar orbit	d
Meteor-M	Russia	2009	Inclination = 98.7°; altitude = 826 km	e
Elektro-L	Russia	2011	Geostationary (76°E)	f

[a] http://isro.org/ (click on Satellites and then click Geo-Stationary Satellites)
[b] http://www.jma.go.jp/jma/jma-eng/satellite
[c] http://www.esa.int/esaEO/SEM9NO2VQUD_index_0_m.html
[d] http://www.sinodefence.com/space/satellite/fenyun3.asp
[e] http://www.n2yo.com/satellite/?s=35865
[f] http://www.russianspaceweb.com/2011.html

Table 3.7. Weather satellites operated by NOAA, an agency of the United States

Satellite name	Year of launch	Orbit	Location	Source
GOES-11	2000	Geostationary	135°W, 0°N	a
GOES-12	2001	Geostationary	60°W, 0°N	a
GOES-13	2006	Geostationary	75°W, 0°N	a
GOES-14	2009	Geostationary	105°W, 0°N	a
GOES-15	2010	Geostationary	90°W, 0°N	a
NOAA-15	1998	Altitude = 807 km Inclination = 98.5°	–	b
NOAA-16	2000	Altitude = 849 km Inclination = 99.0°	–	b
NOAA-17	2002	Altitude = 810 km Inclination = 98.7°	–	b
NOAA-18	2005	Altitude = 854 km Inclination = 98.7°	–	b
NOAA-19	1998	Altitude = 870 km Inclination = 98.7°	–	b
METOP-A	2006	Altitude = 817 km Inclination = 98.7°	–	b

[a] http://www.oso.noaa.gov/goesstatus/
[b] http://www.oso.noaa.gov/poesstatus/

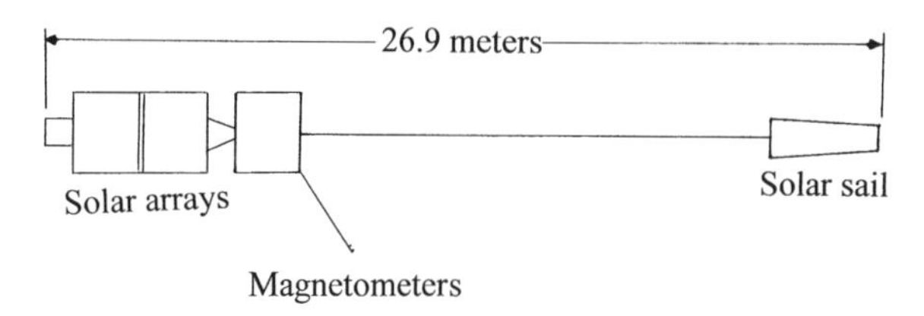

Fig. 3.21. Main parts and dimension of the GOES weather satellite. This data is from Space Systems-Loral, Reference #S-415-19 (GOES I-M Data Book), which is on the website http://goes.gsfc.nasa.gov/text/goes.databook.html (Credit: Richard Schmude, Jr.).

power outage or a gasoline shortage. High energy particles and radiation may damage satellites and harm astronauts.

The energetic particles and X-ray sensors, X-ray imager and magnetometer on *GOES* all collect data on space weather near Earth. The energetic particles sensor detects fast-moving electrons, protons and alpha particles. An alpha particle has two protons and two neutrons. The X-ray sensor monitors the solar flux at wavelengths of 0.05–0.3 nm and 0.1–0.8 nm. It also images our Sun. These images give astronomers insights about developing solar flares and other events that affect space weather. Each pixel has a resolution of 5 arc seconds. The magnetometer measures the strength and direction of the local magnetic field. It is at the end of a 3-m pole. Even at this distance, magnetic fields from the satellite are significant and should be subtracted to yield the correct magnetic field intensity.

Chapter 4

Observing Artificial Satellites

Introduction

This portion of the book deals with the observation and imaging of satellites. It contains three chapters. This chapter describes how to observe artificial satellites. The next one describes images of satellites and research projects involving satellite observations. Chapter 6 gives a mathematical description of how one may compute eclipse times of objects in a geostationary orbit.

Artificial objects in space range in brightness from that of a crescent Moon down to the faintest objects imaged at professional observatories. Many satellites are bright enough to be seen with the unaided eye. These include the International Space Station, the Hubble Space Telescope, some pieces of space junk and many of the satellites described in Chap. 3. Space junk includes broken satellites, spent rocket boosters and fragments from rockets and satellites. Other satellites such as those in geostationary orbit will almost always require optical aid.

We will describe two classes of satellites, those that have low orbits and those with geostationary orbits. Those in low orbits have altitudes below 1,600 km. Those in geostationary orbits have altitudes of around 35,800 km.

This chapter is broken into five major sections. The first one describes how and where to look for low-orbit and geostationary satellites. We will also explore the geometry of satellites, a useful website and the author's sky brightness measurements. The second section gives a description of how to observe these objects. The brightness and standard magnitude of a satellite is described in the third. In the last two sections, you will learn about the author's observations of low-orbit and geostationary satellites.

Looking for Satellites

Unlike stars and planets, the right ascension and declination of a satellite may change by several degrees in just a few minutes. For this reason, different techniques should be used to look for these objects. You will learn how to identify and look for low-orbit and geostationary satellites in this section.

R. Schmude, Jr., *Artificial Satellites and How to Observe Them*, Astronomers' Observing Guides, DOI 10.1007/978-1-4614-3915-8_4, © Springer Science+Business Media New York 2012

Low-Orbit Satellites

There are several unique characteristics of low-orbit satellites. Most of them move 0.4–1.0°/s or the length of a fully extended fist in 10–25 s. They usually remain visible for at least 30 s. They drift silently across the sky and do not leave a vapor trail. Satellites following nearly polar orbits often have a nearly constant brightness. Those moving from west to east (or east to west) will grow brighter or dimmer as a result of a changing phase. A few also tumble and, hence, undergo rapid brightness changes. One may confuse a plane, shooting star, balloon, bright star or planet for a satellite. Characteristics of these objects are summarized in Table 4.1. Let's review objects that may be confused for satellites.

The jet plane is a common object in many areas. In most cases it has one or more flashing red lights. They flash about once a second. Jet planes almost always have two or more lights that may be resolved with the eye or binoculars. All satellites,

Table 4.1. Characteristics of objects that may be confused with artificial satellites

Planes
Red light blinks about once a second
Two or more lights are visible through binoculars
Noise is often associated with planes
Vapor trail is often associated with planes
"Shooting stars"
Usually visible for less than 2 s and move at least 10°/s
More frequent during meteor showers
Diffuse luminous objects[a]
Usually visible for less than 2 s and move at least 10°/s
Appear wider than a normal meteor streak
Balloons
Move slowly
Usually not visible at night
Bright star or planet
No noticeable movement
Brightness is nearly constant
Low-orbit satellites
Moves the length of a fully extended fist in 10–25 s
Visible as one point of light
They do not: flash once a second, leave a vapor trail or have sound
Geostationary satellites
Visible as one point of light
Remains at a constant position; does not rise or set like the stars
They do not: flash once a second, leave a vapor trail or have sound

[a]See Gallagher (1992)

however, appear as a single point of light. In a few cases, such as when a jet is moving directly towards you, it appears as a single bright object. After a few seconds, its direction changes and the flashing red lights appear.

One can mistake a bright star or planet for a satellite during twilight. This is because almost no stars are visible at this time. The best approach in this situation is to keep track of it for at least 30 s. If the suspected object is a satellite, it should move the length of one to three fully extended fists. It is also helpful to know the positions of the brightest stars and planets during twilight. The magazine *Sky & Telescope* is an excellent source for identifying these objects. Another device that may be useful is a planisphere. It shows approximate positions of stars and planets at any time of the year.

Shooting stars and balloons may be mistaken for satellites. Shooting stars streak across the sky. They typically move at least 20°/s and are visible for less than 2 s. You might typically observe a bright circular object at around sunset. Upon closer inspection with binoculars, it may turn out to be a balloon. A satellite, on the other hand, is too far away to show a shape through binoculars. One may also distinguish between a balloon and a satellite by looking for movement. Over a period of 30 s, a satellite should move at least 10° whereas a distant balloon will hardly move.

Geostationary Satellites

Geostationary satellites are dimmer than many of those in low orbits. This is because they are farther away. They move in tiny ellipses high above Earth. See Fig. 4.1. Since they move just a little, they remain at nearly the same position at

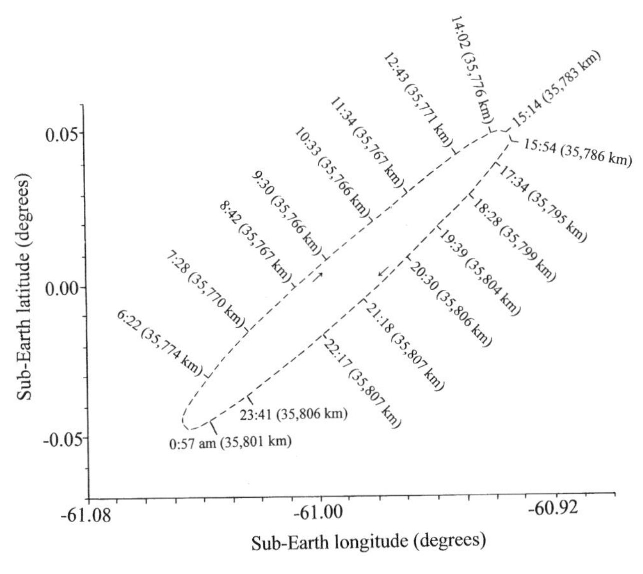

Fig. 4.1. The positions of the sub-Earth point of *Amazonas 2* on October 10, 2011. The sub-Earth point is the location on Earth's surface where this satellite is at zenith. Altitudes are in *parentheses*. Information was taken from http://www.n2yo.com/satellites/?c=10 (Credit: Richard Schmude, Jr.).

Observing Artificial Satellites

all times. One may observe them with a telescope having a field-of-view of about 1.0°. This is twice the angular size of the Moon. When you find a geostationary satellite in the telescope it will remain in nearly the same position with respect to the eyepiece field. At this point, turn off the telescope drive. Stars near it will drift by.

These satellites often change in brightness. In one case, *Echostar 15* reached the threshold of unaided-eye visibility (magnitude 6). This was on October 4, 2011, at 2:22 UT. After about 15 min, its brightness fell by a factor of four. Specular reflection probably caused this brightening.

Geostationary satellite eclipses occur about 1% of the time. These events may yield valuable data on Earth's atmosphere. Eclipses are discussed in Chaps. 5 and 6.

When to Look for Satellites

One may observe geostationary satellites at any time during the night. They can even be observed with nearby streetlights and a waxing gibbous Moon. In the case of viewing low-orbit satellites with the unaided eye, the situation is different. Dark skies are needed. In this section, you will first learn about the geometry for viewing satellites. Next, you will learn about a useful website and sky brightness. Afterwards, you will learn when to observe low-orbit and geostationary satellites.

Geometry for Viewing Satellites

Like the Moon, artificial satellites reflect light. They do this in one of two modes, namely diffuse and specular. Both modes are described in Chap. 1. In most cases, diffuse reflection is dominant. Because of this, we will first learn about the observing geometry and satellite phase.

A low-orbit satellite is best seen during the night. Therefore, it should be high enough to be in sunlight but, at the same time, the observer should be in darkness. See Fig. 4.2. In this drawing, the sky is dark because the observer lies far enough into Earth's umbra. At the same time, the satellite is above Earth's shadow and, hence, reflects light. The dashed line in Fig. 4.3 is the observer's horizon. The Sun is 20° below the horizon and, hence, its altitude is −20°. In this situation, the observer would see a dark sky.

Satellites are brightest when they have a large phase and are at the zenith. Atmospheric extinction will cause objects near the horizon to grow dim. Figure 4.4 shows the brightness drop (in stellar magnitudes) for an elevation of 250 m and for Los Alamos, New Mexico (elevation 2,200 m). You should look at least 30° above the horizon near sea level. For those at a high altitude, look at least 15° above the horizon. The satellite phase is also critical. In the late evening, look to the east. Before sunrise, look to the west. This is because objects are brightest when they are nearly opposite from the Sun. At this position their phase is larger. The phase is the fraction of an object that sunlight illuminates as seen by the observer. See Fig. 4.5. A satellite with a large phase will be much brighter than one with a small phase.

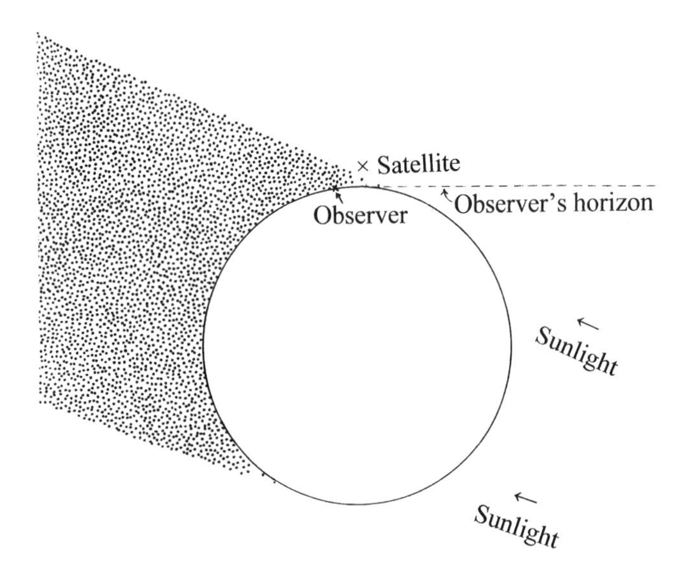

Fig. 4.2. The observer is in Earth's shadow. The satellite, on the other hand, is above Earth's shadow. As a result, the observer is able to see the satellite reflect light (Credit: Richard Schmude, Jr.).

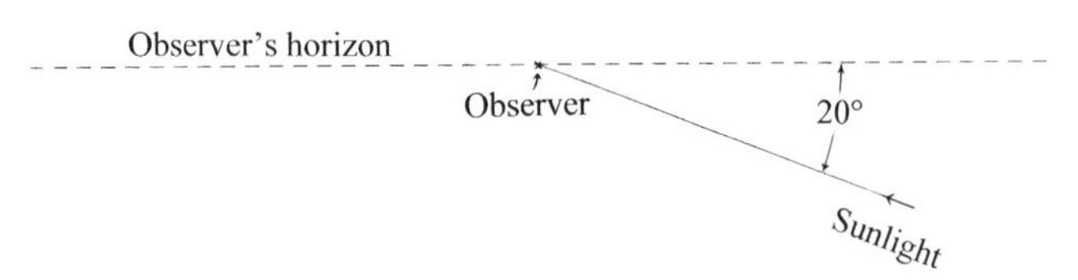

Fig. 4.3. The Sun is 20° below the horizon and, hence, the observer lies in Earth's shadow and is in darkness. The Sun's elevation in this drawing is −20° (Credit: Richard Schmude, Jr.).

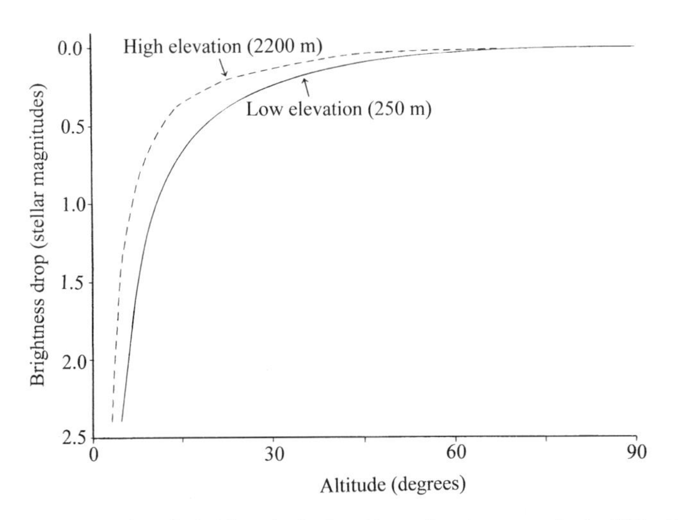

Fig. 4.4. The brightness drop of a satellite for different altitudes. The *solid curve* is for my location at an elevation of 250 m. The *dashed curve* is for an observing site at an altitude of 2,200 m. Both curves are based on the author's measurements of the V-filter extinction coefficient (Credit: Richard Schmude, Jr.).

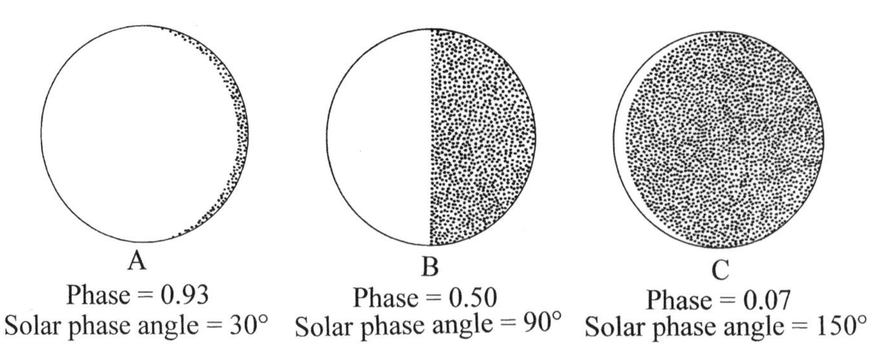

Fig. 4.5. The relationship between the phase and the solar phase angle for a spherical object. As the solar phase angle increases, the phase decreases. The solar phase angle is the angle between the Sun and the observer measured from the object. The phase is the fraction illuminated as seen by the observer (Credit: Richard Schmude, Jr.).

In each of the three cases in Fig. 4.5, sunlight illuminates half of the three-dimensional sphere. The difference between the drawings is the location of the observer in relation to the direction of sunlight. In Fig. 4.6, the observer is in three different locations and, hence, sees three different phases. In the top drawing, the solar phase angle, which is the angle between the observer and the Sun measured from the object's center, is 30°. Therefore, the observer sees a phase of 0.93. When the solar phase angle is 90° the observer sees a phase of 0.50, and when it is 150° the observer sees a phase of 0.07. As the solar phase angle increases, the phase decreases. Our Moon behaves in the same way. It is brightest at full phase. This is because the entire lit side of the Moon faces us and the dark half faces away. Figure 4.7 shows the relative V-filter brightness of the Moon from new to full phase. The Moon is over 100 times brighter with a phase of 0.93 than with one of 0.07. In the same way, a satellite is brighter at a full than at a new phase.

A Useful Website

One website, http://www.heavens-above.com/?, lists predicted positions for satellites each night. In order to properly use it, the location of the observing site should be entered correctly. Use the "edit manually" key in the configuration section near the top of the page. Once the edit manually button is clicked, the program will ask for five characteristics of the observing site, namely latitude, longitude, elevation above sea level, location name and time zone. You will learn about each of these.

The latitude should be entered to the nearest 0.01°. One may use a Global Positioning device or a program like the free site: http://www.itouchmap.com/latlong.html to determine latitude. One should also enter it in decimal degrees. If the latitude is given in degrees, minutes and seconds it should be converted to degrees. Latitudes south of the equator are negative, whereas those north of it are positive.

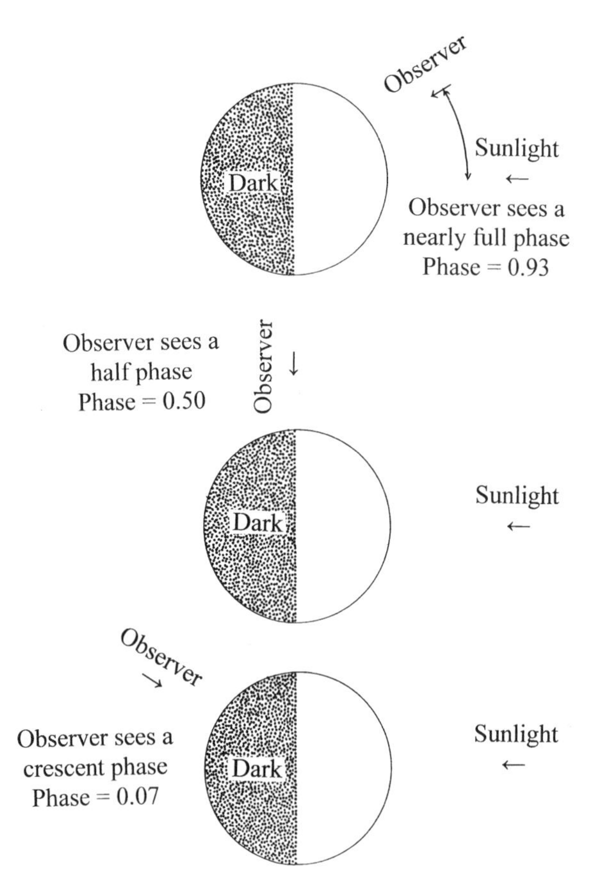

Fig. 4.6. The Sun illuminates about half of a spherical object. The phase, however, depends on the observer's position. The phase increases as the angle between the observer and the sunlight decreases (Credit: Richard Schmude, Jr.).

Like latitude, the longitude of the observing site should be entered. It is entered to the nearest 0.01° in decimal degrees. All longitudes west of the Prime Meridian are negative, whereas those east of it are positive. The continental United States is west of the Prime Meridian and, hence, longitudes are negative.

The elevation of the observing site is entered in meters above sea level. Elevations may be obtained from Google Earth or from an atlas. One should divide an elevation given in feet by 3.28 to convert it to meters.

The location of the observing site should be entered. This is not critical. If the site does not have a name, type in the nearest town or post office.

The time zone should be selected from a list. Without this information, the times will make no sense. The program automatically corrects for both Daylight Savings Time and British Summer Time. Once all information is entered, one may obtain predictions on many types of satellites including the International Space Station, Iridium flares and the Hubble Space Telescope.

The printout gives several pieces of information that include the name of the satellite, the time and location of it at maximum altitude and its brightness in stellar magnitudes. Let's discuss brightness and location.

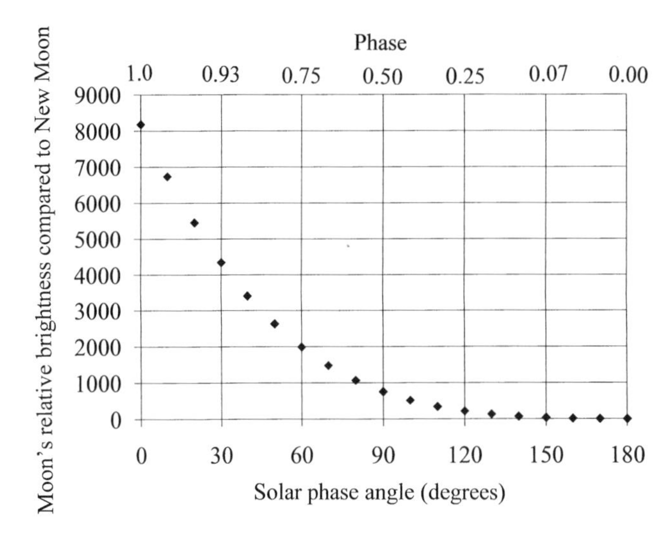

Fig. 4.7. This graph shows the relative brightness of the Moon from new to full phase. At a phase of 0.93, the relative brightness is 4,300, whereas it is less than 40 at a phase of 0.07. These results are based on the author's V-filter measurements. See *Journal of the Royal Astronomical Society*, volume 95, p. 20, 2001. The solar phase angle is also shown on the *bottom axis* (Credit: Richard Schmude, Jr.).

The brightness of stars is measured in stellar magnitudes. Bright stars like Vega and Arcturus have a magnitude of zero, whereas those barely visible to the unaided eye in a dark sky have a magnitude of 6. The larger the star's magnitude value, the fainter it is. Essentially, a star of magnitude 3.0 is 2.512 times brighter than one of magnitude 4.0.

The satellite's position is given in the altitude-azimuth system instead of the right ascension-declination system. One device that may yield accurate altitudes is the quadrant. Figure 4.8 shows a drawing of how to put it together. One attaches a protractor to a block of wood and then attaches string to the protractor's center. The other end of the string has a weight. A 0.25 m piece of ½ in. PVC pipe is placed on top. The string and weight always point straight down. Therefore, when the quadrant is held level, one should see the horizon and the string should line up with 0°. When one aims the quadrant at the zenith, the string should line up with 90°. Figure 4.9 shows a quadrant.

The azimuth describes the direction. Azimuth is measured from 0° up to 360°. The respective azimuth values of north, east, south and west are 0°, 90°, 180° and 270°. Figure 4.10 shows abbreviations for different directions. For example, NNW stands for north northwest, and it is three-fourths the way from west to north. Its azimuth value is 337.5°. One may estimate the azimuth using a compass, the location of the North Star or binoculars with a compass. One should correct for magnetic declination when using a compass. This is because a compass needle points to the magnetic pole instead of true north. Figure 4.11 illustrates magnetic declination. For the eastern half of the United States, the compass needle points a little to the left of true north. For the western half of the country, the needle points in the other direction. One may determine the magnetic declination for their location at http://www.magnetic-declination.com/.

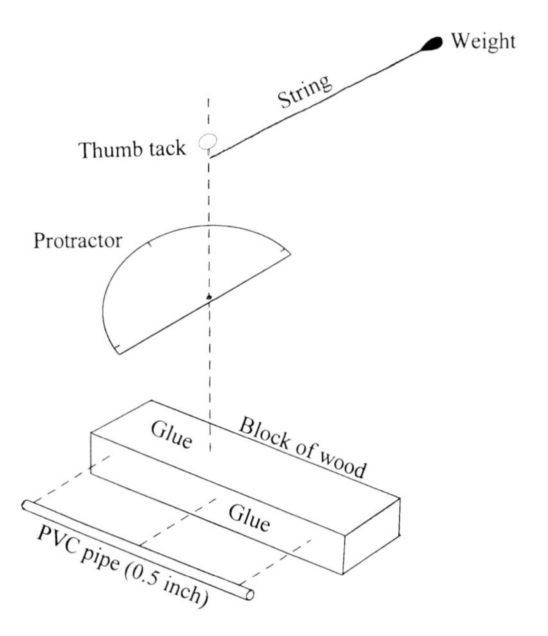

Fig. 4.8. This drawing shows the major parts of a quadrant. The protractor and PVC pipe are glued or taped to the wood block (Credit: Richard Schmude, Jr.).

Fig. 4.9. This image shows a quadrant aimed at the horizon. The altitude equals the angle that lines up with the string (Credit: Richard Schmude, Jr.).

The best way to estimate a satellite's location is to find a bright object near the same altitude as the predicted satellite position. You use a quadrant to measure altitudes of potential objects until a suitable one is found. You may use the North Star or a compass to get the correct direction.

Sky Brightness Measurements

The sky brightness is critical for observing satellites with the unaided eye. More satellites will be visible in a darker sky than in a brighter one. For this reason, we need to learn about the sky brightness measurements.

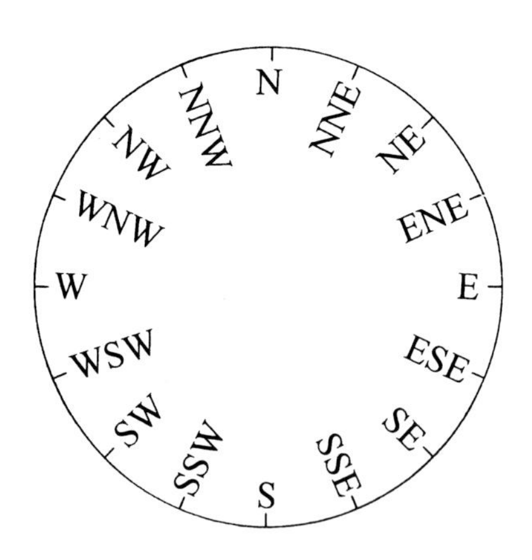

Fig. 4.10. Locations of abbreviated directions for satellites listed in http://www.heavens-above.clom/? (Credit: Richard Schmude, Jr.).

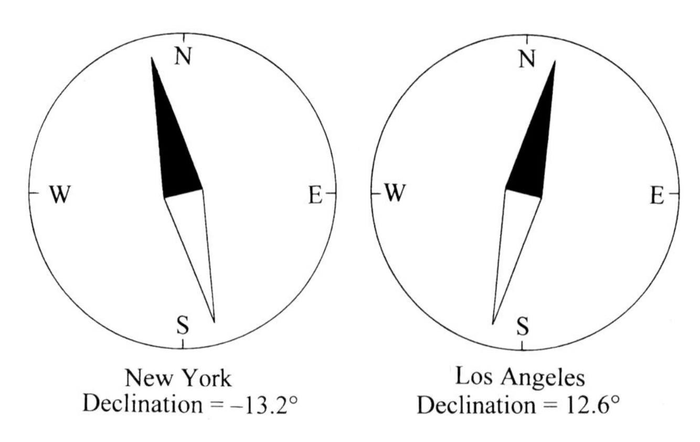

Fig. 4.11. In both compasses, N points to true north. In New York, the compass needle points 13.2° to the left of true north. In Los Angeles, the compass needle points 12.6° to the right of true north. These values are for 2011. The magnetic declination changes, and, hence, observers should check it for their location every 4 or 5 years. The magnetic dedination values are from http://www.magnetic-declination.com/ (Credit: Richard Schmude, Jr.).

Figure 4.12 shows sky brightness measurements, at zenith, versus local time. These were made during the evening of January 5–6, 2010, with a "Sky Quality Meter." The sky began getting dark before sunset. It grew darker as the Sun sank below the horizon. By 19:00 (7:00 pm) local time the sky was as dark as it got for the night. Figure 4.13 shows a graph of sky brightness versus the Sun elevation. At sunset, the sky brightness was about magnitude 7.5 per square arc second. This is equivalent to the light coming from a 7.5 magnitude star spread over each square arc second of the sky. When the Sun was 14° below the horizon (−14°) the sky brightness had dropped to magnitude 19.7 per square arc second. It remained near this value until the next morning. Therefore, once the Sun elevation drops to −14°, conditions are optimum for satellite observation.

During which month is the moonless sky darkest? To answer this question, the author measured the sky brightness 170 times from his backyard between August 2009 and July 2011. All measurements were made at zenith. The Moon was probably below the horizon for all measurements. The monthly average sky brightness values are plotted in Fig. 4.14. For my backyard, the sky has nearly the same brightness throughout the year. Therefore, all months in my area have nearly the same sky brightness.

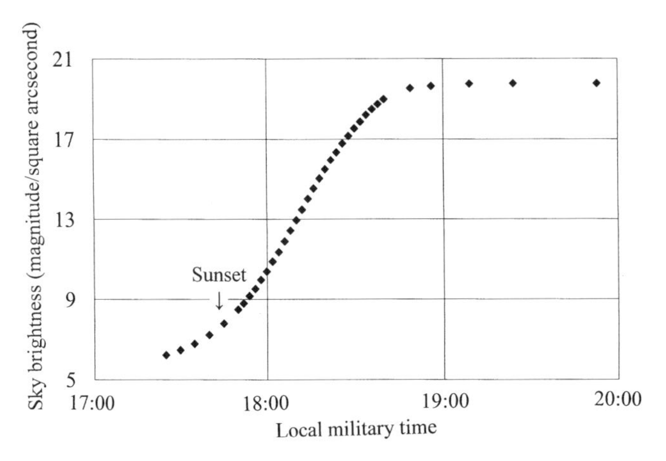

Fig. 4.12. The sky brightness at zenith is plotted against the local military time from Barnesville, Georgia. This data was recorded on the evening on January 5, 2010. The sky brightness unit, magnitude/arc sec, expresses how dark the sky is. For example, a sky brightness of 19.0 means that each square arc second has a brightness of a magnitude 19.0 star (Credit: Richard Schmude, Jr.).

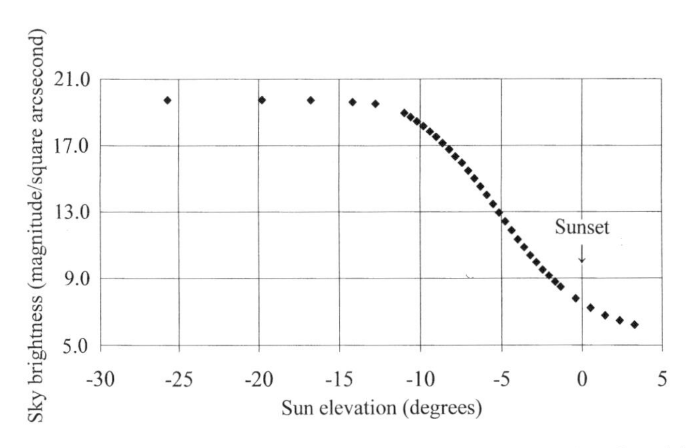

Fig. 4.13. The sky brightness in Fig. 4.12 is plotted against the Sun's elevation. As the Sun gets lower (lower elevation) the sky gets darker. Once the Sun reaches an altitude of −15°, the sky brightness levels off for the night (Credit: Richard Schmude, Jr.).

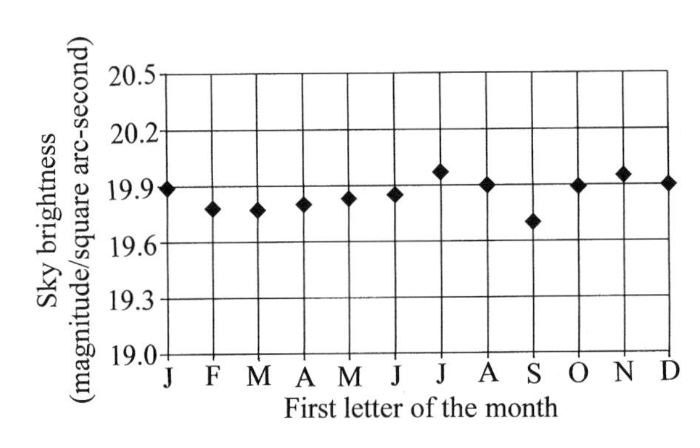

Fig. 4.14. Monthly average sky brightness values based on measurements between August 2009 and July 2011. All measurements were taken in Barnesville, Georgia (Credit: Richard Schmude, Jr.).

Best Seeing for Low-Orbit Satellites

Low-orbit satellites are best seen at large phases under dark skies. Figure 4.15 shows the number of low-orbit satellites predicted to be visible between July 1 and 10, 2011. All data are for Barnesville, Georgia. All predictions are based on data from the website: http://www.heavens-above.com/? The writer believes the data in Fig. 4.15 show a typical distribution for summer months in the continental United States. According to this figure, the largest number of satellites is visible between 22:00 and 23:30 and between 4:00 and 5:30 local time. The times depend on the season. A better indicator of satellite visibility is the Sun's altitude. Most satellites are visible when the Sun is less than 26° below the horizon. Accordingly, one will have the best chance of seeing them when the Sun is between 14° and 26° below the horizon. The lower limit in this range is from sky brightness and the upper limit is from the trend in Fig. 4.15.

Table 4.2 shows the best times to observe low-orbit satellites for different latitudes. It lists the number of minutes after sunset (or before sunrise) when the Sun is between 14° and 26° below the horizon. To use the table, one would look up his or her latitude and select the closest one. Look up the time of sunset for your area and add the values in Table 4.2 to determine the best time interval to look for them. The values are subtracted from the time of sunrise for early morning viewing. For example, the Sun sets at 8:40 pm on July 16 in Atlanta, Georgia. Since Atlanta is at 34° N, one uses the 34° column and matches it up with the mid-July row. The best time in the evening is 56–164 min after 8:40 pm or 9:36 pm to 11:24 pm. As a second example, an observer in Austin, Texas, wants to observe before sunrise during mid-July. The Sun rises at 6:40 am. Since Austin is at 30° N, one would use the 30° column. The best time to observe is 149–53 min before sunrise or from 4:11 am to 5:47 am. For latitudes not listed in Table 4.2, select the one closest to where you live.

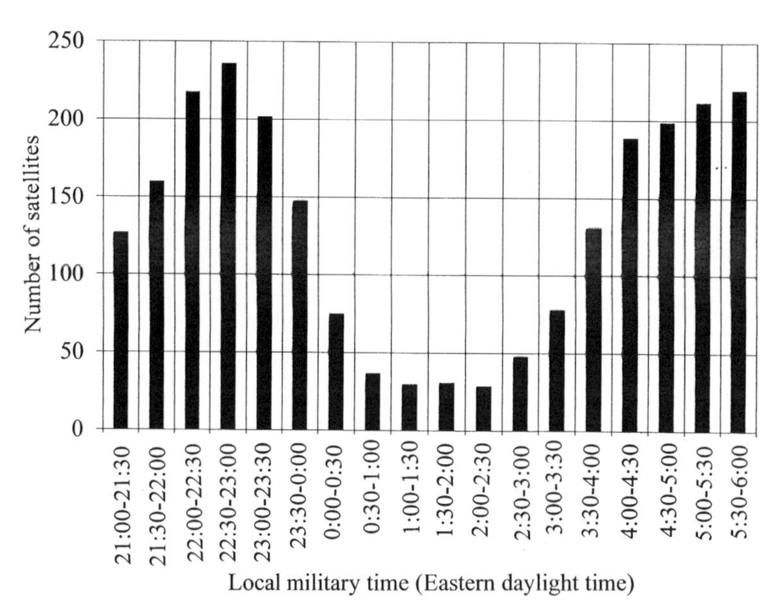

Fig. 4.15. The number of satellites predicted to be visible from Barnesville, Georgia, between 21:00 and 6:00. This data is based on predictions at http://www.heavens-above.com/? (Credit: Richard Schmude, Jr.).

Table 4.2. Best time to observe satellites (minutes after sunset or before sunrise)

	18° N	22° N	26° N	30° N	34° N	38° N	42° N	46° N	50° N	−54° N
Mid-January	45–115	46–118	48–122	50–126	52–132	55–139	59–147	64–158	70–172	79–190
Mid-February	43–111	44–114	46–117	47–121	49–127	52–133	55–142	59–152	64–164	71–180
Mid-March	42–110	43–113	45–116	46–121	48–127	51–134	54–143	58–154	62–169	68–188
Mid-April	43–114	44–117	46–122	48–129	50–137	53–147	57–160	61–179	67–209	75–289
Mid-May	45–122	47–127	49–134	51–143	54–155	58–174	63–204	70[a]	79[a]	94[a]
Mid-June	47–128	49–134	51–143	54–155	57–173	62–203	69[a]	78[a]	92[a]	119[a]
Mid-July	46–125	48–131	50–139	53–149	56–164	60–188	66–241	74[a]	86[a]	106[a]
Mid-August	44–116	45–121	47–126	49–134	52–143	55–156	59–173	64–201	71–276	81[a]
Mid-September	42–110	43–114	45–118	46–123	49–129	51–137	54–148	58–161	63–178	70–203
Mid-October	42–110	44–113	45–116	47–121	49–126	51–133	54–141	58–151	63–164	69–180
Mid-November	44–114	45–116	47–120	49–124	51–130	54–137	58–145	62–156	68–169	76–186
Mid-December	46–117	47–120	49–123	51–128	53–134	56–141	61–150	66–162	73–177	82–196

[a]Observations may be carried out unfel dawn since the sunset elevations remains above −26°

Best Seeing for Geostationary Satellites

Geostationary satellites are dim objects. Therefore, they are best seen in the dark sky. For most of the year, they do not move into Earth's shadow. On dates within about 23–24 days of equinox (March or September), they move into Earth's shadow at a certain time. Essentially, they are in eclipse. This is described in Chaps. 5 and 6. The best time to observe these objects is when they are nearly opposite from our Sun. Therefore, on dates that are more than 23–24 days from equinox, select a satellite opposite from our Sun. Look for an object that has a right ascension either 12 h higher or 12 h lower than that of our Sun. When observing near equinox, select a target that has a right ascension either 11 h higher or 11 h lower than that of our Sun.

Observing Iridium Flares

An Iridium flare is a fascinating event to watch. It occurs when a satellite mirror reflects light to the observer. Essentially, specular reflection occurs. In some cases, the satellite will get as bright as magnitude −8 or −9, which is much brighter than the International Space Station. Therefore, even though the ISS has over 100 times the surface area of an Iridium satellite, the latter may be much brighter.

Figure 4.16 shows a small portion of middle Georgia. The dashed line represents an area where the specular reflection of *Iridium 65* was predicted to be greatest on July 7, 2011. Essentially, this is where the mirror angle equals 0.0°. The closer one is to the dashed line; the brighter will be the predicted flash. Table 4.3 lists the predicted flash characteristics for positions A through F. In Fig. 4.16, point B has the smallest "distance to the flare center" value. Consequently, the flare was predicted to be brighter at that point than at the others. As is shown, a difference of a few kilometers can make a big difference in the observed brightness. Other predicted flashes are like the event illustrated in this drawing.

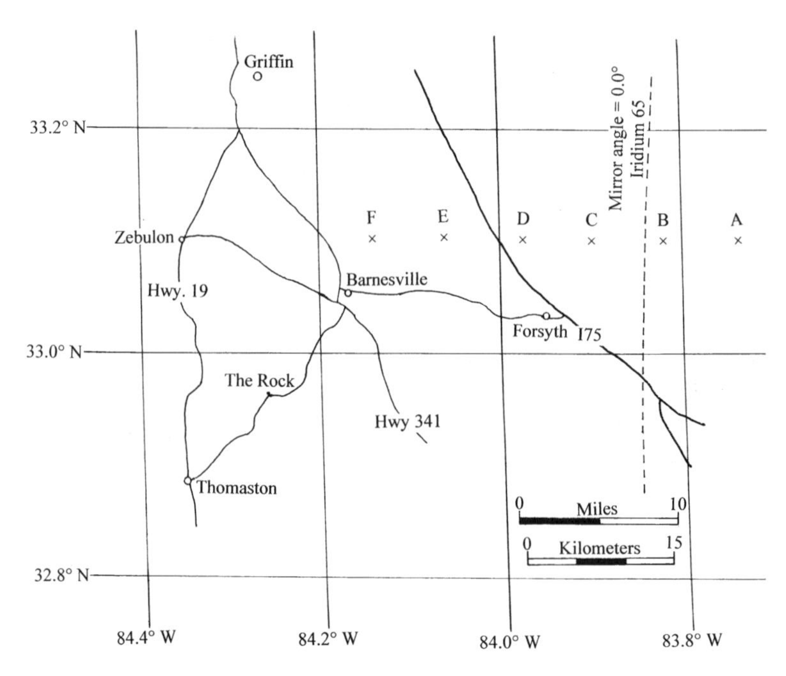

Fig. 4.16. A map of a small portion of middle Georgia. The *dashed line*, to the right, represents the path of maximum brightness of the *Iridium 65* flash. As one gets closer to this *dashed line*, the brighter will be the flash (Credit: Richard Schmude, Jr.).

Table 4.3. Characteristics of the *Iridium 65* satellite on the evening of July 7, 2011, from middle Georgia (Data are from http://www.heavens-above.com/?)

Point in Fig. 4.16	Distance from line where mirror angle = 0.0 (km)	Brightness (magnitudes)	Mirror angle (degrees)
A	10	−4	0.6
B	2	−8	0.1
C	5	−6	0.3
D	13	−4	0.7
E	20	−2	1.2
F	28	−1	1.6

On the website http://www.heavens-above.com/?, the predicted time and position of Iridium flares are given. Positions are given in terms of altitude and azimuth. The "Distance to Flare Center" is also listed. This is the shortest distance between the imputed coordinates and the path of maximum brightness.

Satellite Standard Magnitude

According to Taylor (1990) the standard magnitude (or intrinsic brightness) of a satellite is its brightness, in stellar magnitudes, at a distance of 1,000 km at half phase. Since Earth is at an average distance of 1.0 AU (au) from our Sun, the

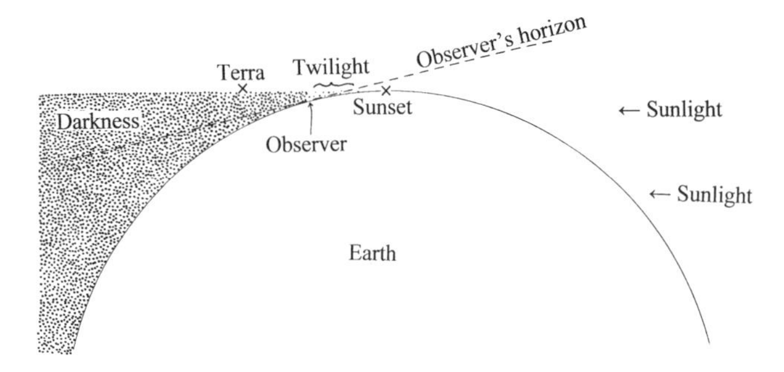

Fig. 4.17. The geometry of the *Terra* satellite at the low solar phase angle of 9°. Note that the observer has just moved into *darkness*, and *Terra* is just above Earth's shadow. Earth is rotating counterclockwise in this illustration (Credit: Richard Schmude, Jr.).

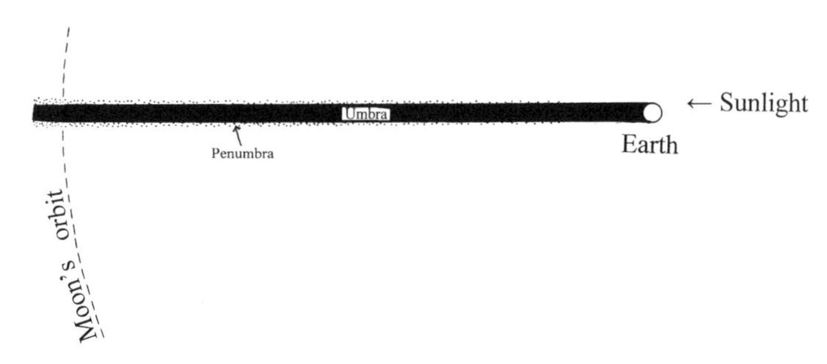

Fig. 4.18. The geometry of the Moon's orbit and Earth's shadow. When the Moon is just outside Earth's shadow, its solar phase angle is 2° and its phase is 0.9997 (Credit: Richard Schmude, Jr.).

satellite-Sun distance should also be 1.0 au for the standard magnitude. Satellites are like our Moon. As discussed earlier, the closer a satellite is to full phase, the brighter it will be.

Artificial satellites are much closer than our Moon. For example, the *Terra* satellite is only 700 km above Earth's surface, whereas the corresponding value for our Moon is 380,000 km. The maximum phase that *Terra* may reach depends on several factors. The phase reaches 1.00 at sunset and at a satellite altitude of 0°. These are unrealistic observing conditions. If one is in darkness (Sun altitude of −14°) and *Terra's* altitude is 23° its phase would equal 0.99. See Fig. 4.17. The Moon's phase, on the other hand, may reach 0.9997 in total darkness before it enters Earth's shadow. The geometry of the Moon's orbit and Earth's shadow is shown in Fig. 4.18.

A satellite's brightness depends on several factors, including albedo, distance, size and standard magnitude. The albedo expresses the fraction of light that an object reflects. For example fresh snow reflects a greater percentage of light than a piece of coal. Therefore, snow has a higher albedo than coal. A satellite with a high albedo will be brighter than one with a low albedo. The closer and larger a satellite is, the more light it will reflect. Finally, a satellite with a bright standard magnitude will usually be brighter than one with a dim standard magnitude. Table 4.4 lists the standard magnitudes of a few satellites. The values in Table 4.4 do not include spurious specular reflection that may occur. In the case of the ISS, brightness values are based

Table 4.4. Standard magnitude of selected satellites. The magnitude values are from the website http://www.heavens-above.com/?

Satellite	Standard magnitude or intrinsic brightness (stellar magnitudes)	Maximum brightness at perigee and at phase = 1.00 (stellar magnitudes)
Aqua	4.0	2.5
Cosmos 1869	4.7	2.2
Envisat	3.7	2.5
Hubble Space Telescope	2.2	−0.3
International Space Station	−1.3	−5.5
Landsat 5	4.7	3.2
NOAA 19	3.6	2.9
SeaSat 1	3.2	1.9
Spot 5	3.7	2.9
Swift	3.7	1.3
Terra	2.7	1.2

on an orbit with a ground distance of 380 km. The altitude of the ISS changes because of gas drag and periodic boosts. For example, according to http://www.heavens-above.com/?, the altitude of the ISS was 346 km on May 1, 2011, but after a boost in June, its altitude was 389 km on July 1, 2011.

Brightness of Amazonas 2

In this section, some measurements of *Amazonas 2* made on October 15 from 2:16 to 3:28 Universal Time will be given. Equation 4.1 shows the relationship between the brightness and the solar phase angle (α) for this satellite.

$$\text{Vvis}(\text{stellar magnitudes}) = 7.0 + 0.11\alpha \qquad [10.2^\circ < \alpha < 25.5^\circ] \qquad (4.1)$$

In this equation, Vvis is the brightness of *Amazonas 2* based on eye estimates and α is the solar phase angle in degrees. This equation is only valid for solar phase angles between 10.2° and 25.5°. When α drops below 9°, Earth will eclipse it. **Therefore, this satellite attains a brightness of 8th magnitude just before it enters Earth's shadow. It grows about one stellar magnitude dimmer per 9° increase of α.** The large value of 0.11 before α indicates that Eq. 4.1 describes specular reflection. In the case of diffuse reflection, the term before α would be close to 0.01. This is described further at the end of Chap. 5.

Brightness of Low-Orbit Satellites

Low-orbit satellites have a wide range of solar phase angles. This undoubtedly leads to a wide range of brightness values. Figure 4.19 shows the distribution of predicted satellite brightness values for July 1 through 10, 2011. Only those equal to or brighter than magnitude 4.5 are included. Most of them are fainter than magnitude 3.5. There is a chance that several of these have grown dimmer over time. This is described further in Chap. 5.

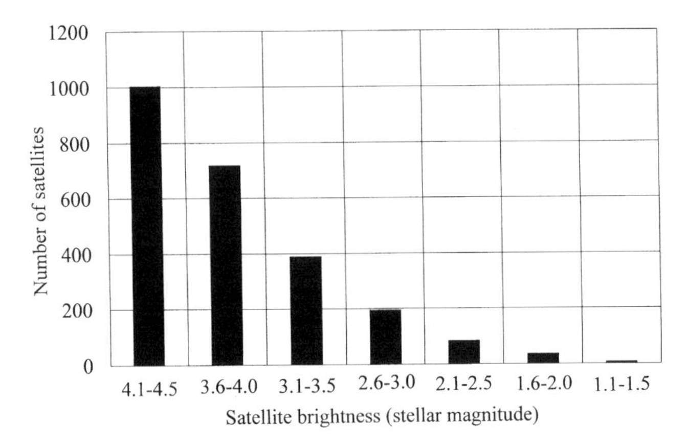

Fig. 4.19. The number of satellites at different brightness levels. All satellites predicted to be visible from Barnesville, Georgia, between July 1 and 10, 2011, are included (Credit: Richard Schmude, Jr.).

Observing Low-Orbit Satellites

Here we will describe an experiment that was carried out involving satellite visibility. The goal was to determine how many satellites one should see along with the best time for observing them. Predicted satellite positions were not consulted. Over a period of 38.2 h, 178 satellites were observed with the unaided eye. Results are summarized in Fig. 4.20.

The upper left graph in the figure shows the number of satellites observed for different Sun altitudes. In all cases the Sun was below the horizon. Since some observations were cut off for a variety of reasons a normalization procedure was used. Essentially, First the Sun altitude for each observed satellite was computed. The satellites were grouped into different categories of the Sun altitude. Afterwards, something called "the number of degrees" was computed. So, if observations were carried out over 10 days and the sky was observed at all times when the Sun had an elevation between −14° and −16° the number of degrees would be 10 × 2° = 20° for the −14° to −16° interval. If, on the other hand, observations were made on just 3 days when the Sun had an altitude between −26° and −28° the total number of degrees would be 3 × 2° = 6° for the −26° to −28° interval. The number of satellites for each Sun elevation interval was divided by the number of degrees. The end results were plotted. According to this figure, more satellites are seen when the Sun is between 12° and 20° below the horizon than for other Sun positions.

The graph in the upper right shows the direction of satellite movement. The observed satellites are placed into one of four groups: North to South, South to North, East to West and West to East. The most common movement was South to North. The least common was East to West. This is because it is harder to launch a satellite in this direction. Essentially, one is working against the rotation of Earth. This is discussed further in Fig. 1.17 and the related text in Chap. 1.

The peak brightness of satellites in stellar magnitudes was also estimated. The results are shown in the graph in the lower left. Most satellites observed had a brightness of between magnitude 2 and 4. On most of the dates, the limiting magnitude was around 5.5. Therefore, fainter satellites were not included.

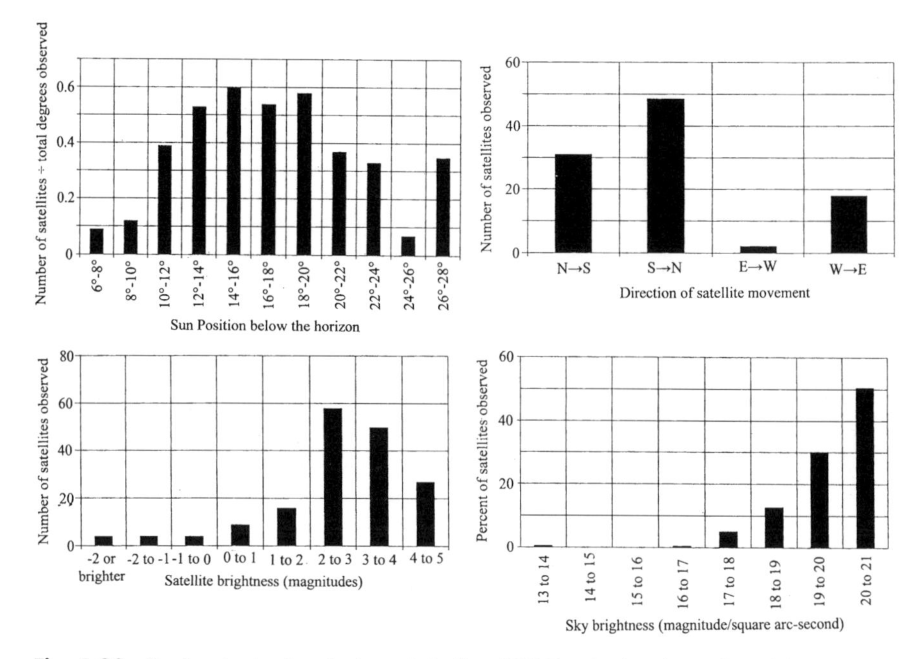

Fig. 4.20. All results are based on observations between April and June of 2011. The author observed 178 satellites with the unaided eye for 38.2 h. During this time, predicted times and locations of satellites were not used. Results are in four graphs. The *upper left graph* shows the number of satellites observed divided by the total degrees observed. The total degrees observed had to be used since observing was stopped early on some dates. For example, if you were to observe for 5 days and watch the sky the entire time when the Sun is 14–16° below the horizon, the total degrees would equal $5° \times 2° = 10°$. The *upper right* graph shows the distribution of satellites moving in the four different directions. The *lower left* graph shows the number of satellites observed versus peak brightness. The *lower right* graph shows the percentage of satellites first detected for different values of the sky brightness (Credit: Richard Schmude, Jr.).

When is the best time to observe satellites with no knowledge of when and where to look? To help answer this question, you can measure the sky brightness with a Sky Quality Meter during the first few seconds after initially seeing a satellite. The percentage of satellites observed for different sky brightness values are plotted in the lower right of Fig. 4.20. The darkest sky for the data in the figure was magnitude 20.7 per square arc second. Over half of the satellites observed had a sky brightness value fainter than magnitude 20.0 per square arc second. Therefore, to answer the question of when to look, the best times are illustrated in Table 4.2.

Observations of Geostationary Satellites

Geostationary satellites require different observing techniques. As mentioned earlier, they move at the same rate as Earth's rotation. They do not rise or set. One should use a telescope or large binoculars to observe them. A clock drive should not be used. Table 4.5 lists a few of them along with their standard magnitude

Table 4.5. Standard and maximum brightness of geostationary satellites. The longitudes and standard magnitudes are from: http://www.arachnoid.com/satfinderonline/index.php and http://www.calsky.com/es.cgi. The author computed the typical brightness for a distance of 35,800 km at half phase. All satellites were launched in 2006 or later except for *Echostar 4*

Satellite	Longitude	Standard magnitude or intrinsic brightness (stellar magnitudes)	Brightness near maximum phase (stellar magnitudes)
Turksat 3a	42.0° E	~4	~9
Hotbird 8	13.1° E	~4	~9
Hotbird 9	13.0° E	~4	~9
Hotbird10	7.3° W	~4	~9
Intelsat 25	31.5° W	3.0	8
Amazonas 1	−61.0° W	5.4	10.6
Amazonas 2	−61.0° W	4.1	7.8
Echostar 12	−61.3° W	5.4	9.9
Echostar 3	−61.4° W	5.4	9.1
Echostar 15	61.5° W	4.1	10.8
Horizons 2	74.0° W	3.2	8
GOES 13	74.9° W	3.4	8
Echostar 4	76.8° W	4.7	10
Echostar 11	110.2° W	3.0	8
Echostar 14	118.9° W	3.0	8

and other characteristics. Many are visible from the continental United States, Alaska and Hawaii. *Turksat 3a, Hotbird 8–10* and *INTELSAT 25* are visible from the United Kingdom.

Predicting the brightness of a satellite is like predicting how an American football will bounce. This is because satellites have irregular shapes. Furthermore, specular reflection may be a large percentage of reflected light; this is not the case for the Moon or planets. Because of this, it is difficult to predict the brightness of a satellite. The author has estimated the brightness of five geostationary satellites at a solar phase angle of 10–20°. The satellites, with respective brightness values in stellar magnitudes in parentheses, are: *Amazonas 1* (10.6), *Amazonas 2* (7.8), *Echostar 12* (9.9), *Echostar 3* (9.1) and *Echostar 15* (10.8). The average brightness change between the standard magnitude and the one just before eclipse is 4.8 stellar magnitudes, with a standard deviation of 1.8 stellar magnitudes. Essentially, this factor accounts for distance, solar phase angle and changes in specular reflection. Therefore, a satellite with a standard magnitude of 5.4 may reach a brightness of magnitude 10.2 just before it enters Earth's penumbra. The uncertainty in this value is two stellar magnitudes.

A geostationary satellite does not have a fixed right ascension. Therefore, one should know where it will pass ahead of time. The website http://www.n2yo.com/satellites/?c=10 lists the current right ascension (RA) and declination (δ) of many satellites. It takes approximately 60 min of time for a geostationary satellite to move 1 h of RA. A procedure for finding one is illustrated in Fig. 4.21. We will look at an example. Let's say you want to observe *Amazonas 2* at around 11 P.M in October. You would first look up its right ascension (RA) and declination. Let's say

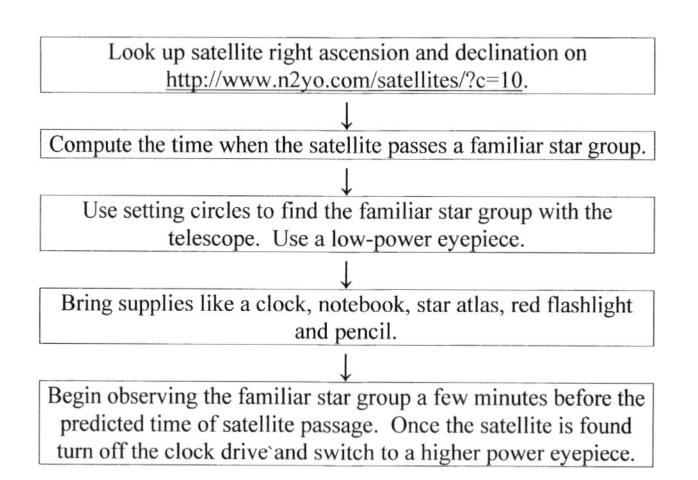

Fig. 4.21. This figure illustrates a procedure for searching for geostationary satellites (Credit: Richard Schmude, Jr.).

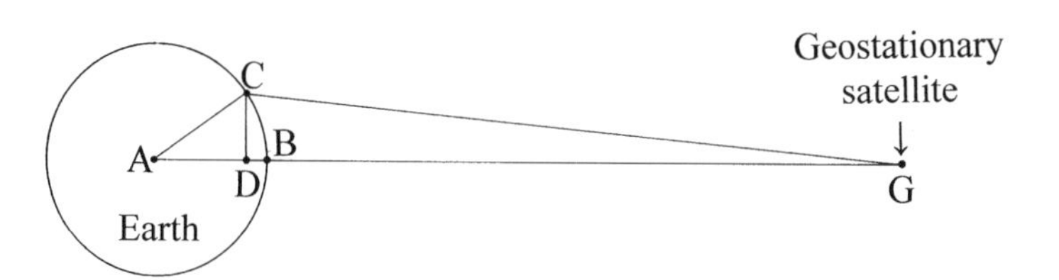

Fig. 4.22. The drawing shows the geometry of Earth and the declination of a geostationary satellite. If the observer is at point C and in the northern hemisphere, the declination would equal angle CGD. Furthermore the satellite would be south of the celestial equator (Credit: Richard Schmude, Jr.).

at 10:00 P.M. it is RA = 23 h 10 m and δ = 5.38° S. Let's say that you are familiar with a group of stars near RA = 0 h 01 m and δ = −5.5°. Therefore, the target will move across this group at 10:51 pm. Once this is computed, you set up the telescope and aim it on this group. You then insert a low-power eyepiece and begin looking for the target a few minutes before 10:51 P.M. Once you site *Amazonas 2*, you insert a higher-power eyepiece. This darkens the sky.

On some nights, you may have trouble with condensation on the eyepiece. If this happens, place it near a heater, and this will eliminate the problem.

Although geostationary satellites lie close to Earth's equatorial plane, they do not have a declination of 0°. This is because they are closer than 45,000 km. See Fig. 4.22. An observer north of the equator will observe the satellite south of the celestial equator. The opposite is true for an observer in the southern hemisphere. Table 4.6 lists the declination of geostationary satellites for different observer latitudes.

Figure 4.23 shows Earth's shadow and nomenclature for a satellite eclipse. Earth has an angular diameter of almost 18° for geostationary satellites. In the penumbra, Earth blocks part of the Sun and in the umbra, it blocks all of the Sun. The same nomenclature used for lunar eclipses is used here. In Fig. 4.23 point P1 is the point where the satellite first enters the penumbra. Since this object is less than 30 m long, it takes less than 0.01 s for it to move into the penumbra. Therefore, P2 is at almost the same time as P1. Point U1 is the point when the satellite enters the umbra and U4 is when it leaves the umbra. Point P4 is the point when it leaves

Table 4.6. The declinations of a geostationary satellite for different observer latitudes. It is assumed the satellite lies in Earth's equatorial plane. The data are based on the author's calculations

Latitude	Declination	Latitude	Declination
63° N	−8.22	30° N	−4.96
60° N	−8.05	27° N	−4.53
57° N	−7.86	24° N	−4.08
54° N	−7.64	20° N	−3.44
51° N	−7.39	15° N	−2.62
48° N	−7.12	10° N	−1.76
45° N	−6.82	30° S	4.96
42° N	−6.49	33° S	5.38
39° N	−6.15	36° S	5.77
36° N	−5.77	39° S	6.15
33° N	−5.38	42° S	6.49

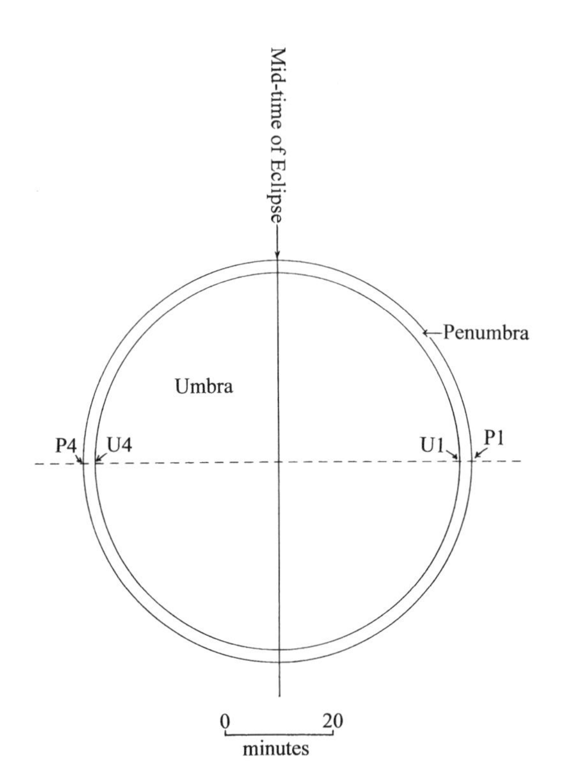

Fig. 4.23. This drawing illustrates the penumbra and umbra portions of Earth's shadow for a geostationary satellite. The satellite moves from right to left along the *dashed line*. Point P1 is location where the satellite enters the penumbra, and P4 is the location where it exits the penumbra. Point U1 is the location where the satellite enters the umbra, and U4 is the location where it exits the umbra (Credit: Richard Schmude, Jr.).

the penumbra. The length of time between P1 and P4 is up to 72 min. A scale representing the time is given in Fig. 4.23.

Watching a satellite pass into Earth's shadow is a unique event. It starts out with the satellite maintaining a nearly constant brightness. It starts to get dim once it is about half-way into the penumbra. Afterwards, it dims rapidly. Once the satellite reaches the umbra it continues to fade without a sharp drop-off. Once it is just inside the umbra, it may disappear and reappear. This behavior is due to our atmosphere bending light towards and away from the satellite. The data is valuable because it may help us better understand our atmosphere. At some point it becomes invisible. The point that it becomes invisible depends on the limiting magnitude. As an example, during one of the author's eclipse observations, the faintest stars visible were near magnitude 12. Therefore, the target had fallen below this value once it was deep enough into Earth's shadow.

How will the brightness of a geostationary satellite change as it moves into Earth's shadow? To answer this question, let's look at lunar eclipses. Let's say you measured the brightness of a small area on the Moon during the January 20–21, 2000, eclipse. This area's brightness fell four stellar magnitudes as it moved through Earth's penumbra. It dimmed an additional 4.5 stellar magnitudes from U1 to a depth of 1,000 km in the umbra. Therefore, the total brightness drop was 8.5 stellar magnitudes. Under similar circumstances, a geosynchronous satellite may have a stellar magnitude of 7.5 before it enters Earth's penumbra. If it behaves in the same way as the Moon did on January 20–21, 2000, it would drop to magnitude 11.5 near U1 and would drop to magnitude 16 once it is 1,000 km into the umbra. Even though *Amazonas 2* entered the umbra, it was usually visible for a few seconds. This is discussed further in Chap. 5. There is one additional uncertainty, though, and that is the role of specular reflection at low solar phase angles.

Geostationary satellites undergo eclipses during two 6.5-week periods each year. These occur between late February and early April and again between late August and mid October. Eclipses happen when the Sun's declination is between 9° N and 9° S. The path it follows depends on the Sun's declination. Figure 4.24 shows the path

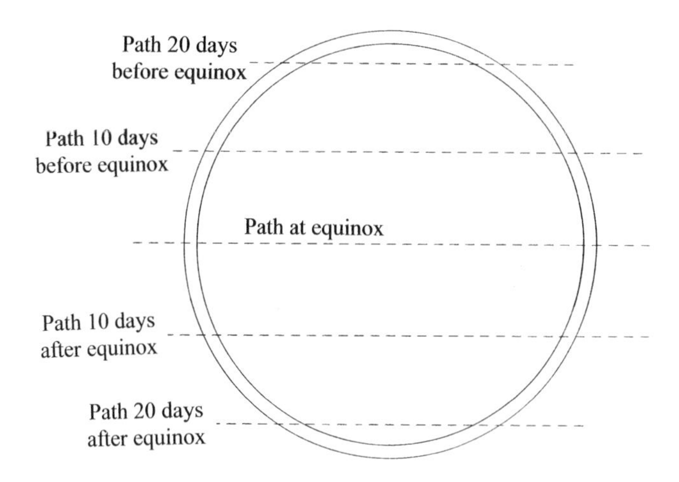

Fig. 4.24. A geostationary satellite will move through the center of Earth's shadow on the day of equinox in September. At other times, it will move above or below the center, as illustrated (Credit: Richard Schmude, Jr.).

Amazonas 2 follows before and after the September equinox. The situation is nearly the same in other years since the cycle of the Sun's declination repeats every year. A similar set of eclipses take place during the March equinox. On a few dates, the satellites in Table 4.6 pass through just the penumbra. Eclipse times depend on the longitude of the satellite and the observer's time zone. Tables 4.7, 4.8, 4.9, 4.10, 4.11, 4.12, 4.13 and 4.14 list predicted times of geostationary satellite eclipses for the United Kingdom and different regions of the United States. These are based on a solid, airless Earth. The method used in computing predicted eclipse times is described in Chap. 6.

Figure 4.25 shows the amount of time a geostationary satellite spends in Earth's umbra and penumbra. Calculations are based on the solid Earth and do not include atmospheric effects. The assumed latitude of all satellites is 0.00°. Eclipses may last up to 72 min near equinox. One may estimate the length of an eclipse by knowing the date and time of equinox. For example, a satellite will remain in the umbra for 60 min on October 5 (12 days after equinox). It will spend 2.5 min in the penumbra in going from P1 to U1 and another 2.5 min in going from U4 to P4. Therefore, it will spend a total of 5 min in the penumbra.

Table 4.7. Predicted times that two geostationary satellites will enter/exit Earth's penumbra (P1/P4) and umbra (U1/U4). All times are computed from the longitudes in the two websites http://www.arachnoid.com/satfinderonline/index.php and http://www.calsky.com/es.cgi. Eastern Daylight Times and military times are listed

Time of observation	P1	U1	U4	P4
Intelsat 25; 31.5°W				
20 days before March equinox	22:01	22:06	22:31	22:36
10 days before March equinox	21:44	21:46	22:46	22:48
March equinox	21:38	21:40	22:47	22:49
10 days after March equinox	21:38	21:41	22:40	22:42
20 days after March equinox	21:49	21:54	22:21	22:25
20 days before September equinox	21:47	21:51	22:20	22:24
10 days before September equinox	21:30	21:32	22:32	22:34
September equinox	21:23	21:25	22:32	22:34
10 days after September equinox	21:23	21:25	22:25	22:27
20 days after September equinox	21:34	21:39	22:06	22:10
Amazonas 2; 61.0°W				
20 days before March equinox	23:59	0:04	0:29	0:34
10 days before March equinox	23:42	23:44	0:44	0:46
March equinox	23:36	23:38	0:45	0:47
10 days after March equinox	23:36	23:39	0:38	0:40
20 days after March equinox	23:47	23:52	0:19	0:23
20 days before September equinox	23:45	23:49	0:18	0:22
10 days before September equinox	23:28	23:30	0:30	0:32
September equinox	23:21	23:23	0:30	0:32
10 days after September equinox	23:21	23:23	0:23	0:25
20 days after September equinox	23:32	23:37	0:04	0:08

Those who observe lunar eclipses report that Earth's umbra is around 2% larger than what is predicted from a solid, airless Earth. This observation is based on the location of maximum contrast between the penumbra and umbra. Essentially, the location of highest contrast differs from that of the umbra boundary for a solid, airless Earth. This is illustrated in Fig. 4.26. Portions of the penumbra and umbra are shown here. On the left, part of the penumbra is on top and part of the umbra is on the bottom for a solid, airless Earth. In this case, no refracted light enters the umbra. Therefore, it is black. No sunlight reaches it. On the right, parts of the penumbra and umbra are shown for Earth with its atmosphere. These drawings have two differences. The first one showing the area of maximum contrast is much higher in the drawing on the right than the one on the left. This is the umbra enlargement and is discussed further in Chap. 5. A second difference between the two drawings is that the umbra is not black in the right one but is black in the left one. This is because our atmosphere refracts some sunlight into the umbra as is shown in the drawing on the right.

The maximum contrast usually corresponds to a maximum brightness drop per unit of time. Visual observations are consistent with a small umbra enlargement. It will be interesting to make accurate brightness measurements of a satellite as it moves through Earth's shadow. Eclipses are discussed further in Chaps. 5 and 6.

Table 4.15 lists a few important websites that are mentioned in this chapter along with the information they provide.

Table 4.8. Predicted times that two geostationary satellites will enter/exit Earth's penumbra (P1/P4) and umbra (U1/U4). All times are computed from the longitudes in the two websites http://www.arachnoid.com/satfinderonline/index.php and http://www.calsky.com/es.cgi. Central Daylight Times and military times are listed

Time of observation	P1	U1	U4	P4
Intelsat 25; 31.5°W				
20 days before March equinox	21:01	21:06	21:31	21:36
10 days before March equinox	20:44	20:46	21:46	21:48
March equinox	20:38	20:40	21:47	21:49
10 days after March equinox	20:38	20:41	21:40	21:42
20 days after March equinox	20:49	20:54	21:21	21:25
20 days before September equinox	20:47	20:51	21:20	21:24
10 days before September equinox	20:30	20:32	21:32	21:34
September equinox	20:23	20:25	21:32	21:34
10 days after September equinox	20:23	20:25	21:25	21:27
20 days after September equinox	20:34	20:39	21:06	21:10
Amazonas 2; 61.0°W				
20 days before March equinox	22:59	23:04	23:29	23:34
10 days before March equinox	22:42	22:44	23:44	23:46
March equinox	22:36	22:38	23:45	23:47
10 days after March equinox	22:36	22:39	23:38	23:40
20 days after March equinox	22:47	22:52	23:19	23:23
20 days before September equinox	22:45	22:49	23:18	23:22
10 days before September equinox	22:28	22:30	23:30	23:32
September equinox	22:21	22:23	23:30	23:32
10 days after September equinox	22:21	22:23	23:23	23:25
20 days after September equinox	22:32	22:37	23:04	23:08

Table 4.9. Predicted times that two geostationary satellites will enter/exit Earth's penumbra (P1/P4) and umbra (U1/U4). All times are computed from the longitudes in the two websites http://www.arachnoid.com/satfinderonline/index.php and http://www.calsky.com/es.cgi. Mountain Daylight Times and military times are listed

Time of observation	P1	U1	U4	P4
Amazonas 2; 61.0°W				
20 days before March equinox	21:59	22:04	22:29	22:34
10 days before March equinox	21:42	21:44	22:44	22:46
March equinox	21:36	21:38	22:45	22:47
10 days after March equinox	21:36	21:39	22:38	22:40
20 days after March equinox	21:47	21:52	22:19	22:23
20 days before September equinox	21:45	21:49	22:18	22:22
10 days before September equinox	21:28	21:30	22:30	22:32
September equinox	21:21	21:23	22:30	22:32
10 days after September equinox	21:21	21:23	22:23	22:25
20 days after September equinox	21:32	21:37	22:04	22:08
Echostar 14; 118.9°W				
20 days before March equinox	1:51	1:56	2:20	2:25
10 days before March equinox	1:34	1:36	2:35	2:38
March equinox	1:27	1:29	2:36	2:39
10 days after March equinox	1:28	1:30	2:30	2:32
20 days after March equinox	1:39	1:44	2:10	2:15
20 days before September equinox	1:36	1:41	2:09	2:14
10 days before September equinox	1:19	1:22	2:21	2:24
September equinox	1:12	1:14	2:22	2:24
10 days after September equinox	1:12	1:15	2:14	2:17
20 days after September equinox	1:24	1:28	1:55	2:00

Table 4.10. Predicted times that two geostationary satellites will enter/exit Earth's penumbra (P1/P4) and umbra (U1/U4). All times are computed from the longitudes in the two websites http://www.arachnoid.com/satfinderonline/index.php and http://www.calsky.com/es.cgi. Pacific Daylight Times (or Mountain Standard Times) and military times are listed

Time of observation	P1	U1	U4	P4
Amazonas 2; 61.0°W				
20 days before March equinox	20:59	21:04	21:29	21:34
10 days before March equinox	20:42	20:44	21:44	21:46
March equinox	20:36	20:38	21:45	21:47
10 days after March equinox	20:36	20:39	21:38	21:40
20 days after March equinox	20:47	20:52	21:19	21:23
20 days before September equinox	20:45	20:49	21:18	21:22
10 days before September equinox	20:28	20:30	21:30	21:32
September equinox	20:21	20:23	21:30	21:32
10 days after September equinox	20:21	20:23	21:23	21:25
20 days after September equinox	20:32	20:37	21:04	21:08
Echostar 14; 118.9°W				
20 days before March equinox	0:51	0:56	1:20	1:25
10 days before March equinox	0:34	0:36	1:35	1:38
March equinox	0:27	0:29	1:36	1:39
10 days after March equinox	0:28	0:30	1:30	1:32
20 days after March equinox	0:39	0:44	1:10	1:15
20 days before September equinox	0:36	0:41	1:09	1:14
10 days before September equinox	0:19	0:22	1:21	1:24
September equinox	0:12	0:14	1:22	1:24
10 days after September equinox	0:12	0:15	1:14	1:17
20 days after September equinox	0:24	0:28	0:55	1:00

Table 4.11. Predicted times that *Echostar 14,* a geostationary satellite, will enter/exit Earth's penumbra (P1/P4) and umbra (U1/U4). All times are computed from the longitudes in the two websites http://www.arachnoid.com/satfinderonline/index.php and http://www.calsky.com/es.cgi. All times are Daylight Savings Times for most of Alaska including Anchorage. Military times are listed

Time of observation	P1	U1	U4	P4
Echostar 14; 118.9°W				
20 days before March equinox	23:51	23:56	0:20	0:25
10 days before March equinox	23:34	23:36	0:35	0:38
March equinox	23:27	23:29	0:36	0:39
10 days after March equinox	23:28	23:30	0:30	0:32
20 days after March equinox	23:39	23:44	0:10	0:15
20 days before September equinox	23:36	23:41	0:09	0:14
10 days before September equinox	23:19	23:22	0:21	0:24
September equinox	23:12	23:14	0:22	0:24
10 days after September equinox	23:12	23:15	0:14	0:17
20 days after September equinox	23:24	23:28	23:55	0:00

Table 4.12. Predicted times that *Echostar 14,* a geostationary satellite, will enter/exit Earth's penumbra (P1/P4) and umbra (U1/U4). All times are computed from the longitudes in the two websites http://www.arachnoid.com/satfinderonline/index.php and http://www.calsky.com/es.cgi. Hawaii Standard Times and military times are listed

Time of observation	P1	U1	U4	P4
Echostar 14; 118.9°W				
20 days before March equinox	21:51	21:56	22:20	22:25
10 days before March equinox	21:34	21:36	22:35	22:38
March equinox	21:27	21:29	22:36	22:39
10 days after March equinox	21:28	21:30	22:30	22:32
20 days after March equinox	21:39	21:44	22:10	22:15
20 days before September equinox	21:36	21:41	22:09	22:14
10 days before September equinox	21:19	21:22	22:21	22:24
September equinox	21:12	21:14	22:22	22:24
10 days after September equinox	21:12	21:15	22:14	22:17
20 days after September equinox	21:24	21:28	21:55	22:00

Table 4.13. Predicted times that two geostationary satellites will enter/exit Earth's penumbra (P1/P4) and umbra (U1/U4). All times are computed from the longitudes in the two websites: http://www.arachnoid.com/satfinderonline/index.php and http://www.calsky.com/es.cgi. Atlantic Daylight Times and military times are listed

Time of observation	P1	U1	U4	P4
Intelsat 25; 31.5°W				
20 days before March equinox	23:01	23:06	23:31	23:36
10 days before March equinox	22:44	22:46	23:46	23:48
March equinox	22:38	22:40	23:47	23:49
10 days after March equinox	22:38	22:41	23:40	23:42
20 days after March equinox	22:49	22:54	23:21	23:25
20 days before September equinox	22:47	22:51	23:20	23:24
10 days before September equinox	22:30	22:32	23:32	23:34
September equinox	22:23	22:25	23:32	23:34
10 days after September equinox	22:23	22:25	23:25	23:27
20 days after September equinox	22:34	22:39	23:06	23:10
Amazonas 2; 61.0°W				
20 days before March equinox	0:59	1:04	1:29	1:34
10 days before March equinox	0:42	0:44	1:44	1:46
March equinox	0:36	0:38	1:45	1:47
10 days after March equinox	0:36	0:39	1:38	1:40
20 days after March equinox	0:47	0:52	1:19	1:23
20 days before September equinox	0:45	0:49	1:18	1:22
10 days before September equinox	0:28	0:30	1:30	1:32
September equinox	0:21	0:23	1:30	1:32
10 days after September equinox	0:21	0:23	1:23	1:25
20 days after September equinox	0:32	0:37	1:04	1:08

Table 4.14. Predicted times that two geostationary satellites will enter or exit Earth's penumbra (P1/P4) and umbra (U1/U4) for those in the United Kingdom. All times are computed from the longitudes in the two websites: http://www.arachnoid.com/satfinderonline/index.php and http://www.calsky.com/es.cgi. British Summer Times and military times are listed

Time of observation	P1	U1	U4	P4
Turksat 3a; 42.0° E				
20 days before March equinox	22:07	22:12	22:37	22:42
10 days before March equinox	21:50	21:52	22:52	22:54
March equinox	21:44	21:46	22:53	22:55
10 days after March equinox	21:44	21:47	22:46	22:48
20 days after March equinox	21:55	22:00	22:27	22:31
20 days before September equinox	21:53	21:57	22:26	22:30
10 days before September equinox	21:36	21:38	22:38	22:40
September equinox	21:29	21:31	22:38	22:40
10 days after September equinox	21:29	21:31	22:31	22:33
20 days after September equinox	21:40	21:45	22:12	22:16
Hotbird 10; 7.3° W				
20 days before March equinox	1:24	1:29	1:54	1:59
10 days before March equinox	1:07	1:10	2:09	2:11
March equinox	1:01	1:03	2:10	2:12
10 days after March equinox	1:01	1:04	2:03	2:06
20 days after March equinox	1:13	1:17	1:44	1:49
20 days before September equinox	1:10	1:14	1:43	1:47
10 days before September equinox	0:53	0:55	1:55	1:57
September equinox	0:46	0:48	1:55	1:57
10 days after September equinox	0:46	0:48	1:48	1:50
20 days after September equinox	0:57	1:02	1:29	1:34

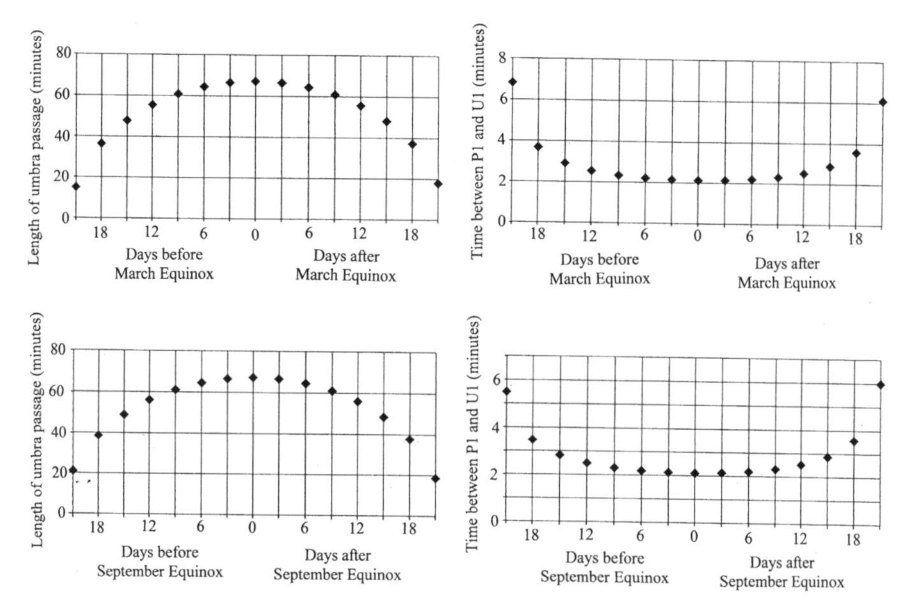

Fig. 4.25. The amount of time a geostationary satellite spends in Earth's umbra (graphs on left) and between points P1 and U1 (graphs on right). Results are based on the procedure outlined in Chap. 6 (Credit: Richard Schmude, Jr.).

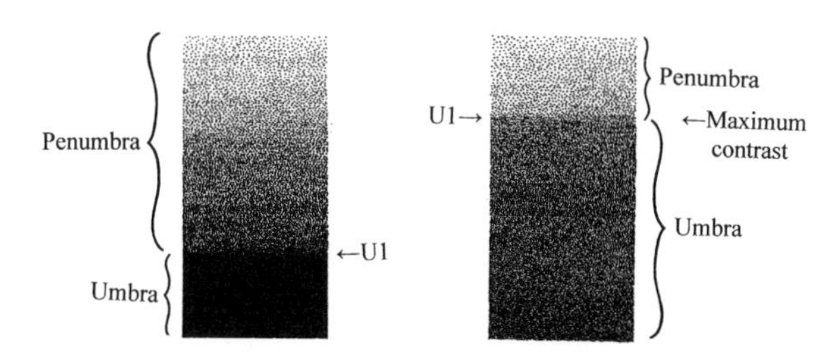

Fig. 4.26. Two drawings are shown that illustrate part of Earth's penumbra and umbra. The drawing on the *left* shows the situation for an airless Earth. Essentially, the border between the penumbra and umbra is very sharp, and no light enters the umbra. The umbra is black. The drawing on the *right* shows the penumbra and umbra for Earth with its atmosphere. Note the area of maximum contrast is higher. Furthermore, the umbra is not totally black. This is because the atmosphere refracts some light that reaches this area (Credit: Richard Schmude, Jr.).

Table 4.15. Websites related to artificial satellites and related information.

Website	Description
http://www.magnetic-declination .com/	Lets you determine the declination of your observing site
http://www.itouchmap.com/latlong.html	Lets you determine the latitude and longitude of your observing site
http://www.heavens-above.com/?	Predicts the positions of hundreds of satellites including Iridium flares and the International Space Station
http://www.arachnoid.com/satfinderonline/index.php	Predicts the altitude and azimuth of geostationary satellites
http://www.calsky.com/es.cgi	Gives the brightness and sizes of many artificial satellites; it also includes finder charts for geostationary satellites
http://www.n2yo.com/satellites/?c=10	Lists the right ascension, declination and altitude of geostationary satellites. It also shows the latitude and longitude of a geostationary satellite's sub-Earth point to an accuracy of 0.01°. Also lists characteristics of low-orbit satellites

Chapter 5

Imaging Artificial Satellites and Doing Research

Imaging Artificial Satellites

Since the late 1990s, astronomers have made great strides in imaging artificial satellites with small telescopes. This is because of the development of sensitive video cameras. Observers have obtained angular resolutions of one-half of an arc second. This corresponds to about a meter for a satellite 400 km away. Others have recorded satellites transiting our Sun and Moon.

An artificial satellite orbiting Earth is different from a planet or deep sky object in two ways. It moves much faster across the sky than planets, and its brightness may change in a few seconds. The rapid brightness change may be caused by a changing distance, a changing phase or a varying amount of specular reflection. These differences pose a special challenge to the operator of a small telescope. This is because it is difficult to obtain hundreds of images to stack. Instead one should be content to capture a few frames of a satellite. In time, it may be possible for mounts to track a satellite. But even if this occurs, one may have to deal with the changing brightness.

In this section, we will explore the work of four individuals who have used different techniques to image satellites. One of these, Ralf Vandebergh, has imaged over a dozen satellites. Others have imaged the International Space Station transiting the Moon, Sun and passing close to Venus.

Tom Faber

Tom uses a Canon Powershot SX100lS camera on a tripod. He usually sets the exposure for 15 s. The camera is focused at infinity and the ISO is selected. Tom also uses a timer that helps reduce unwanted vibrations during an exposure.

Tom made the image in Fig. 5.1. It shows the International Space Station and the space shuttle *Endeavour* flying side by side. The ISS is the brighter streak at the top. It was brightest at the right edge.

R. Schmude, Jr., *Artificial Satellites and How to Observe Them*, Astronomers' Observing Guides,
DOI 10.1007/978-1-4614-3915-8_5, © Springer Science+Business Media New York 2012

Fig. 5.1. This is an image of the International Space Station and the space shuttle *Endeavour*. The ISS is at the top. The exposure was 15 s, and, hence, the fast moving satellites appear as a line (Credit: Tom Faber).

Fig. 5.2. Theo Ramakers and Frank Garner took this image on September 25, 2010, at 20:38:04 Universal Time. It shows Venus and the International Space Station. Essentially, video was taken at a rate of 60 frames per second. Each frame was stacked, showing the resulting image (Credit: Theo Ramakers and Frank Garner).

Theo Ramakers

Theo has been imaging artificial satellites since 2008. He uses a telescope along with a DMK21AU04.AS monochrome Astrovideo camera. The shutter speed is set at 0.001 s, and video is taken at 60 frames per second.

In late 2010, Theo recorded a unique image showing the International Space Station passing Venus. See Fig. 5.2. Below we describe how he constructed this image.

Theo used the CalSky website (listed in Table 4.15) to determine when the International Space Station (ISS) would pass close to Venus. As it turned out, the transit took place in the daytime at 4:38 P.M. Eastern Daylight Time on September 25, 2010. On that day, Theo and Frank Garner set up an 80 mm stellarvue refractor along with a computer, camera and other equipment. They used several steps to make the image in Fig. 5.2. They first recorded several seconds of video of Venus about 1 min before the transit. This video was later used to construct a properly exposed image of Venus. After taking the Venus video, these two adjusted the camera settings to get the correct exposure time for the ISS. Keep in mind the phase of the ISS was like that of Venus and, hence, was dimmer than what it would be at half phase. Shortly afterwards, they recorded the transit of the ISS and Venus. The ISS was correctly exposed but Venus was overexposed in these frames. Theo and Frank used *Photoshop*™ to construct the final image. They constructed a properly exposed image of Venus from the video taken just before the transit. Afterwards they replaced the overexposed Venus area with the correctly exposed image of Venus. This was done for each frame containing the ISS. Therefore, each frame now had both the ISS and Venus correctly exposed. Finally the images were stacked, creating what we see in this figure. Since 60 frames were taken in 1 s, each one shows the ISS in a different location.

Ralf Vandebergh

Ralf has imaged artificial satellites for several years. His technique is summarized in Table 5.1. Ralf uses a well-aligned finderscope with a magnification of 6× to track his target manually. Ralf and his telescope are shown in Fig. 5.3. In many cases he is able to get several images within a second. Since a satellite's appearance usually does not change in a second, he may stack images. After some processing he produces a final image. Figure 5.4 shows an image of the ISS. Parts about 1 m across are visible. Several solar panels are visible along with a few modules. Ralf was able to attain an angular resolution of about one half of an arc second in this image. He also imaged *Meteor 1,* an old Soviet weather satellite. See Fig. 5.5. It was launched in 1969. The image is blurry because the target is small. It is about the size of a sport utility vehicle. In spite of the large distance (425 km or 264 miles), Ralf was able to image its general shape. Figure 5.6 shows the shape of the recently launched *Nanosail D* satellite. This is an experimental solar sail the size of an automobile. Ralf was 667 km (415 miles) from the target when he took the image. Ralf

Table 5.1. The techniques Ralf Vandebergh uses to image artificial satellites

Tracking	Ralf does this manually. He tracks with a 6× finderscope
Method	Eyepiece projection
Eyepiece	15 mm Plossl
Telescope	0.25 m Newtonian
Camera	Video camera (Japan Victor Company)
Processing	Stacks a few frames
Focusing	On a star or other distant object

Fig. 5.3. Ralf Vandebergh and his telescope (Credit: Servé Vaessen).

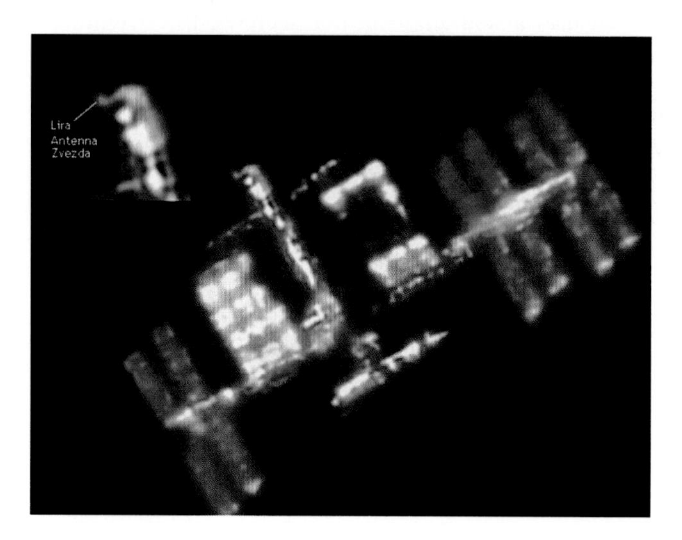

Fig. 5.4. This is an image of the International Space Station. Ralf Vandebergh took it on January 29, 2009, at 17:37:59 UT (Credit: Ralf Vandebergh).

also imaged Mercury and two artificial satellites transiting our Sun. See Fig. 5.7. It was recorded on May 7, 2003. On this date, Mercury subtended an angle of 12.0 arc seconds. Each of the satellites is about one-sixth the angular size of Mercury and, hence, subtended an angle of a couple of arc seconds.

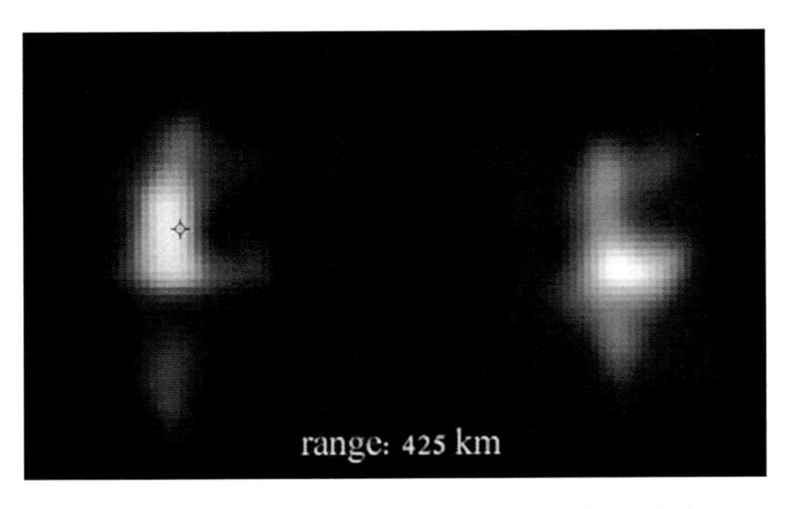

Fig. 5.5. This is an image of *Meteor 1-1*, a weather satellite that the former USSR launched. Ralf Vandebergh took this image on June 2, 2011, at 0:09 UT (Credit: Ralf Vandebergh).

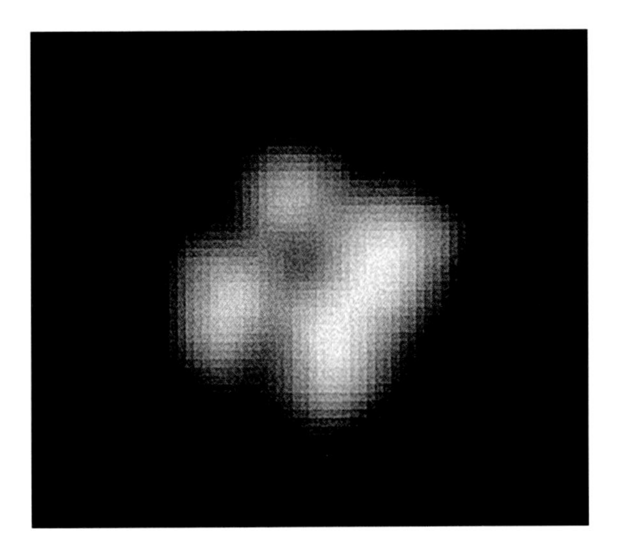

Fig. 5.6. This is an image of *Nanosail-D*, an experimental solar sail designed to test propulsion using the solar wind. *Nanosail-D* has an area of 9 m^2. This is about the size of an automobile. Ralf Vandebergh took this image on May 24, 2011, at 23:20 UT. At that time, the distance to *Nanosail-D* was 667 km (Credit: Ralf Vandebergh).

Ed Morana

Ed modified a program Thomas Fly developed. This program allows one to predict when a satellite will transit our Sun or Moon. One inputs his or her location and the program computes transit times. The ISS Transit Prediction program can be found at Ed's website: http://pictures.ed-morana.com/ISSTransits/predictions/index.html.

Recording a transit can be difficult. One difficulty is the satellite and Moon (or Sun) travel at different angular speeds. An artificial satellite may travel

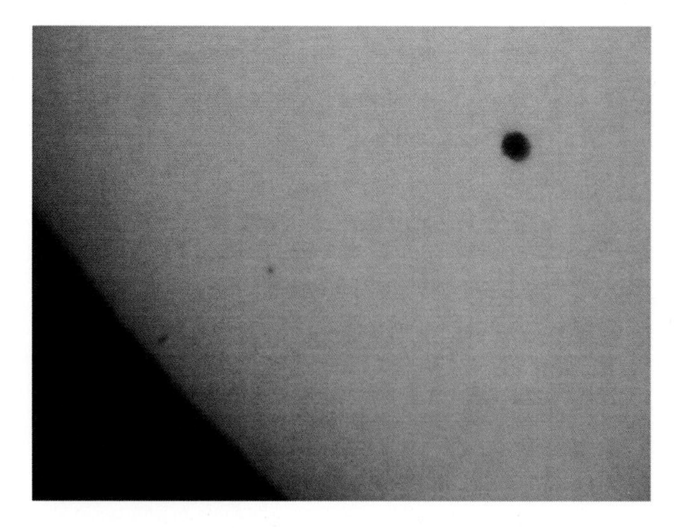

Fig. 5.7. This is an image of Mercury and two artificial satellites transiting our Sun. Ralf took this image on May 7, 2003. At that time Mercury had an angular size of 12.0 arc seconds. Therefore, the satellites had angular sizes of about 2 arc seconds (Credit: Ralf Vandebergh).

Fig. 5.8. An image containing four frames (*eight fields*) showing the ISS transiting our Moon. The ISS moved from right to left. Its range is 383 km. Ed Morana took this image on February 13, 2006, from just outside Newman, California (Credit: Ed Morana).

0.5° per second, but the Moon or Sun may travel 0.004° per second. Because of this difference one should get a single frame of the satellite. A more serious problem is the brightness difference between the Sun (magnitude of −27) and a satellite (magnitude ~0). For this reason, a procedure like the one used to make Fig. 5.2 should be used.

Figure 5.8 shows the ISS transiting the Moon. Ed had to travel about 50 miles to Newman, California, to catch this transit. His technique is summarized in Table 5.2.

Table 5.2. The techniques Ed Morana uses to image ISS transits

Method	Uses Thomas Fly's ISS transit service, obtains latest predictions from Space-Track.org. This allows Ed to focus on a specific part of the Moon where the transit occurs. No stacking, raw frames used. No focal reducer used
Telescope	0.25 m f/10 Schmidt-Cassegrain
Camera	Watec 902H CCD video camera with KIWI OSD video time inserter
Focusing	Focused on the moon

Fig. 5.9. An image containing five frames (*ten fields*) showing the ISS transiting our Sun. The ISS moved from bottom to top. Its distance was 550 km. Ed took this image on August 29, 2006, from Livermore, California (Credit: Ed Morana).

Ed also captured a series of frames showing the ISS transiting our Sun. See Fig. 5.9. He used a solar filter. Other than that, he used the same technique as he did for the Moon.

Research Opportunities and Projects

After observing and imaging artificial satellites, one may wonder about research opportunities involving satellites. What can one learn from satellite observations? Plenty. For example, one may determine how the environment in deep space affects the brightness and color of satellites. One may also monitor Earth's shadow (umbra) and our atmosphere. One would do this by measuring the brightness of a geostationary satellite as it moves through Earth's umbra.

This section will summarize research projects. We will start out with a review of our knowledge of Earth's umbra from lunar eclipses. Afterwards, we will examine eclipse measurements and a few experimental results. The chapter will end with a discussion of brightness measurements of satellites and what we may learn from them.

Umbra Characteristics

The umbra is the area where Earth blocks out the Sun completely. Two sources of sunlight in this area are refracted and scattered light. In refraction, our atmosphere bends sunlight by a small amount. Scattering is different. Sunlight hits the atmosphere and bounces around, and some of it ends up in the umbra. Atmospheric dust forward-scatters a small amount of light. To make matters more complicated, the atmosphere refracts and scatters some colors of light more than others. For example, it refracts more red than blue light. This is why the fully eclipsed Moon often has a reddish hue. The amount of refracted and scattered light depends on several things. A discussion of this is beyond the scope of the book.

Without the atmosphere, an object in the umbra would only be illuminated by starlight and light reflected by Solar System objects. Its brightness would drop about 20 stellar magnitudes. *(This number may be computed in the following way. First compute a combined brightness of magnitude −6.0 for all stars brighter than magnitude 11.50 using information in the Millennium Star Atlas ©1997 by Sinnott and Perrymann, p VII and XI. If the brightness of fainter stars and other objects are included, the combined brightness may reach −7.5. However, half of this light will not reach the side of the object facing us. This reduces the total brightness by 0.75 magnitudes to −6.75. This is 20.0 magnitudes fainter than our Sun.)* Without any sunlight reaching the umbra, the Moon would drop from a brightness of magnitude −13 to magnitude +7. Since the late 1950s, the fully eclipsed Moon has always been at least a factor of 10 brighter than magnitude 7. Therefore, since that time, our atmosphere has always refracted and/or scattered some light into the umbra.

Both the brightness and color of an object changes at it enters the umbra. Therefore, brightness measurements of an object moving through the umbra may yield information on our atmosphere.

Lunar Eclipses: What We Have Learned About Our Umbra

The amount of refracted and scattered light entering Earth's umbra changes from day to day. This is caused by varying amounts of dust and aerosols in the stratosphere and mesosphere along with changing pressure, temperature and chemical composition in the different layers of our atmosphere. Therefore, the Moon's brightness and color will be different from one eclipse to the next. Furthermore, at any given time, some parts of the umbra receive more light than others. Therefore, the brightness of the Moon also changes at different depths in the umbra.

We have a record of lunar eclipses dating back several decades. Because of the extensive record of brightness and color measurements during eclipses, we have a better understanding of Earth's umbra. Many questions, however, remain to be answered. In the next two sections, you will learn about relevant lunar eclipse measurements. Afterwards, you will learn about factors that affect lunar eclipse observations.

One may measure the Danjon number during a total lunar eclipse. Astronomers estimate this on a scale from 0 to 4. A Danjon number of 0 is assigned when the eclipsed Moon is almost invisible and has a dark gray color. On the other hand, a Danjon number of 4 is assigned when it is very bright with a bright copper-red or orange color. Figure 5.10 shows the Danjon number for several total lunar eclipses. Large drops in the Danjon number occurred after major volcanic eruptions in 1963, 1982 and 1991. The data also show the Danjon number returns to a normal value of about 2.0 one to two years after a major eruption. The average Danjon number for all data in the graph is 2.0 with a standard deviation of 0.8.

One may also measure the Moon's brightness drop during a total lunar eclipse. This is computed from

$$\text{Brightness drop}(\text{magnitudes}) = \text{Mb} - [0.19 - 0.35 + 5.0 \times \log(r \times \Delta)] \quad (5.1)$$

In this equation, Mb is the brightness of the eclipsed Moon in stellar magnitudes, r and Δ are the Moon – Sun and Moon – Earth distances, respectively. Both r and Δ are in astronomical units (au). The 0.19 term is from Schmude (2000) and represents the Moon's brightness in stellar magnitudes if it is 1.0 au from Earth and the Sun at full phase. The 0.35 term is the Moon's opposition surge, which is how much it brightens (in stellar magnitudes) when its phase angle approaches zero degrees.

Two ways of measuring Mb are the reverse binocular method and the whole-disc photometry method. In the reverse binocular method, one looks at an object of known brightness in stellar magnitudes with the unaided eye and then looks at the eclipsed Moon through binoculars in reverse. One then estimates its brightness in terms of the object and uses an equation to compute the final brightness value. One may estimate the brightness of the Moon to an accuracy of ~0.5 magnitudes. This method is described in *Comets and How to Observe Them* ©2010 by Richard Schmude, Jr., pp. 130–131. A second way of measuring Mb is the whole-disc photometry method. Since 2000, the writer and co-workers have measured the brightness of the eclipsed Moon using an SSP-3 photometer along with filters transformed to the B, V, R and I system. One may measure Mb values to an accuracy of ~0.07 magnitudes with this method.

Figure 5.11 shows brightness drop values of the eclipsed Moon. The major volcanic eruptions in 1963, 1982 and 1991 caused a larger-than-average brightness drop. Therefore, the umbra was darker than normal. As in the case of the Danjon number, the umbra returned to normal 1–2 years after each eruption. The average

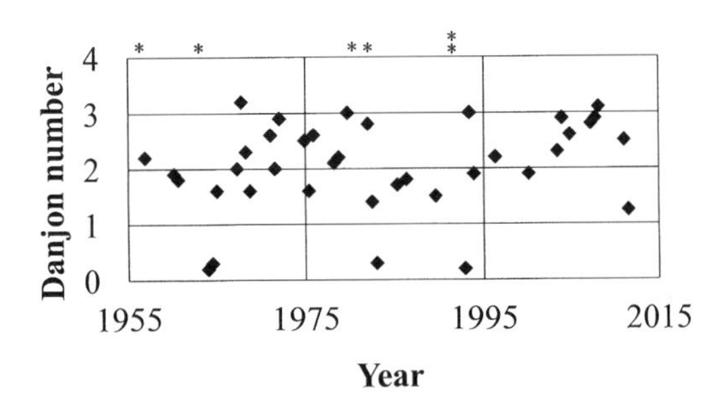

Fig. 5.10. This graph shows the Danjon number of the eclipsed Moon versus the year. The estimated uncertainty of each point is 0.5. Data cover the time period from November 18, 1956, to June 15, 2011. *Asterisks* represent major volcanic eruptions between 1956 and the late 1990s. Danjon number data are from Barbier (1961), Cooper and Geyser (2004), di Cicco (1989), Graham (1987, 1995), Haas (1982), Karkoschka and Aguirre (1996), MacRobert (2007a, b), O'Meara and di Cicco (1993a, 1994b), O'Meara (1993b, 1994a), Rao (2008), Reynolds and Sweetsir (1995), Reynolds and Westfall (2008), Schmude (2004, 2008a), Schmude et al. (2000), Schober and Schroll (1973), Seronik (2000), Sinnott (1996, 2003), *Sky & Telescope* (1964, 1965) and Westfall (1980, 1982, 1986, 1988). Volcanic eruption data are from the sources in Table 5.1 (Credit: Richard Schmude, Jr.).

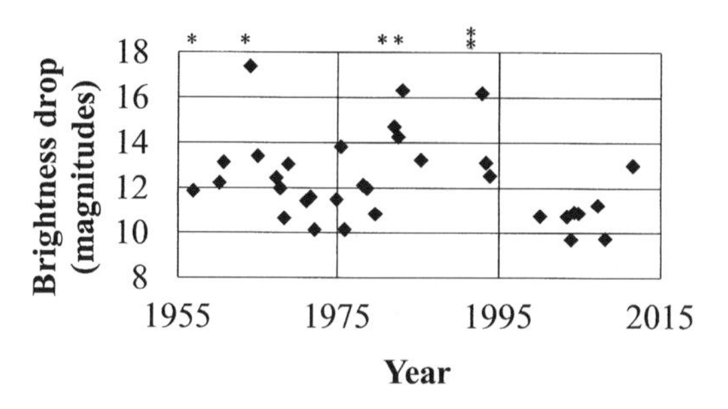

Fig. 5.11. The graph shows the brightness drop of the Moon near mid-totality versus the year. The uncertainty of the January 2000; May 2003 and October 2004 points is 0.1 stellar magnitudes. The uncertainty for all other points is 0.5 stellar magnitudes. The data covers the same time period as in Fig. 5.10. *Asterisks* represent major volcanic eruptions between 1956 and the late 1990s. Data are from Flanders (2005), Haas (1982), O'Meara and di Cicco (1993a, 1994b), O'Meara (1993b; 1994a), Rao (2008), Schmude (2004, 2008a), Schmude et al. (2000), Schober and Schroll (1973), Sinnott (2003, 2004), Sky & Telescope (1964, 1965), Vital (2011), Westfall (1980, 1986, 1988). Volcanic eruption data are from the sources in Table 5.3. (Credit: Richard Schmude, Jr.).

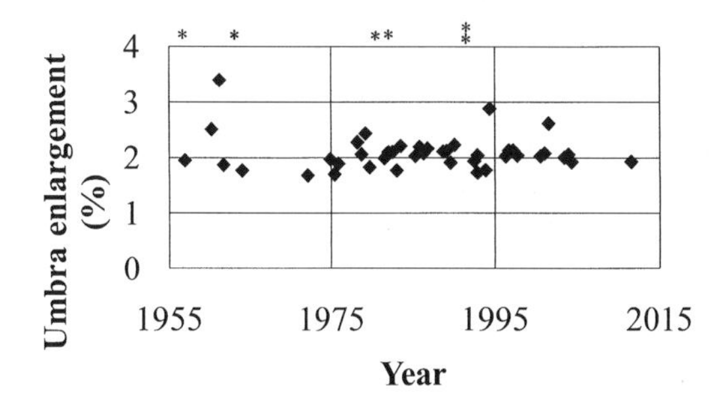

Fig. 5.12. This graph shows the enlargement of Earth's umbra based on crater timings versus the year. A typical uncertainty for each point is 0.1%. Data cover the same period as in Fig. 5.10. *Asterisks* represent major volcanic eruptions between 1956 and the late 1990s. Enlargement values are from Cooper and Geyser (2004), MacRobert (2007b);, Rao (2008), Reynolds and Sweetsir (1995), Reynolds and Westfall (2008), Schmude (2004, 2008a), Schmude et al. (2000), Soulsby (2004), Vital (2011), Westfall (1988). Volcanic eruption data are from the sources in Table 5.3 (Credit: Richard Schmude, Jr.).

brightness drop in Fig. 5.11 is 12.3 stellar magnitudes (or a factor of 86,000). The standard deviation is 1.9 stellar magnitudes.

Finally, one may measure the size of Earth's umbra during a total lunar eclipse. This is more difficult and some explanation is needed.

For a solid body without an atmosphere, the boundary between the penumbra and umbra is sharp. The contrast is highest at this boundary. Earth is different. It has an atmosphere that affects sunlight. Because of the atmosphere, the boundary between the penumbra and umbra is fuzzy. Astronomers have defined this boundary as the point of maximum contrast. See Fig. 4.26. In the next few paragraphs, the umbra is defined in this manner.

Lunar eclipses have enabled astronomers to measure the size of the umbra. During the last several decades, astronomers recorded the times when specific craters passed into the umbra. These have enabled others to compute its size. As it turns out, it is larger than expected. The expected size is based on a solid, airless Earth. The difference between the expected and observed size is expressed as the percent enlargement. A typical value is 2%. Our atmosphere is believed to be the cause of the enlarged umbra. Like the Danjon number and the brightness drop, the percent enlargement changes. Figure 5.12 shows measurements made over the past few decades. The average enlargement is 2.09% with a standard deviation of 0.31%.

Table 5.3. This table lists some of the most violent volcanic eruptions between 1956 and the late 1990s. All of these eruptions except for possibly Agung, had a Volcanic Explosive Index of 5 or higher according to Simkin and Siebert (2000). Bullard (1976) reports Agung released more material into the atmosphere than the Bezymianny eruption and, hence, Agung is included in the table. Some data were also taken from Whitford-Stark (2001). The date of the Agung eruption is from Wikipedia

Volcano	Eruption	Country
Cerro Hudson	1991	Chile
Pinatubo	1991	Philippines
El Chichón	1982	Mexico
Saint Helens	1980	United States
Agung	1963	Indonesia
Bezymianny	1956	Russia

Changes occur from one eclipse to the next. In many cases these are larger than the uncertainties. Let's look at some factors that lead to the changes in Figs. 5.10, 5.11, and 5.12.

The amount of stratospheric ozone affects lunar eclipse measurements. Karkoschka and Aguirre (1996) developed a model that predicts the brightness and color of the eclipsed Moon. For example, in this model, a reduction in stratospheric ozone causes more light to enter the umbra. This is consistent with the reported drop in ozone levels along with my analysis of lunar eclipse data. Essentially, Eq. 5.2 is computed from the linear least squares method.

$$\text{Brightness drop of the Moon}\left(\text{stellar magnitudes}\right) = 12.38 - 0.031\text{t} \qquad (5.2)$$

In this equation, t is the number of years after January 1.0, 1960. Equation 5.2 does not include data from 1956, 1982, 1992 or 1993 due to interference from large volcanic eruptions. This equation shows that the brightness drop of the eclipsed Moon increased by 0.031 magnitude per year between 1960 and 2011. This is consistent with the umbra becoming brighter and, hence, a drop in stratospheric ozone. The correlation coefficient for Eq. 5.2, however, was −0.40. Recall that the correlation coefficient is always between −1.0 and 1.0. Values near zero mean that there is no linear relationship between two variables. A value near 1.0 means that there is a strong linear relationship between the two variables and a value near −1.0 means a strong negative linear relationship between two variables.

Volcanic eruptions also affect lunar eclipse measurements. At least one volcanic eruption occurs each year. It is the large ones on land that affect our atmosphere and climate. Scientists have a number of ways of estimating the strength of these events. One of these is the Volcanic Explosive Index, or VEI. This is based on the volume of material released along with the plume height. The higher the VEI value, the more powerful the eruption. Table 5.3 lists the most powerful events between 1955 and the late 1990s. This is based on the VEI value. These are designated with an asterisk in Figs. 5.10, 5.11, and 5.12. The umbra darkens after major eruptions. No total lunar eclipses took place in 1980 or 1981. Consequently, the impact of the Mount St. Helens event was not measured. By the time of the next lunar eclipse in early 1982, the atmosphere had returned to normal.

Another reason for the changes in lunar eclipses is the umbra itself. The umbra is darkest near the center. For example, Rougier and Dubois (cited in Barbier 1961) carried out brightness measurements of the Moon during the total lunar eclipse of

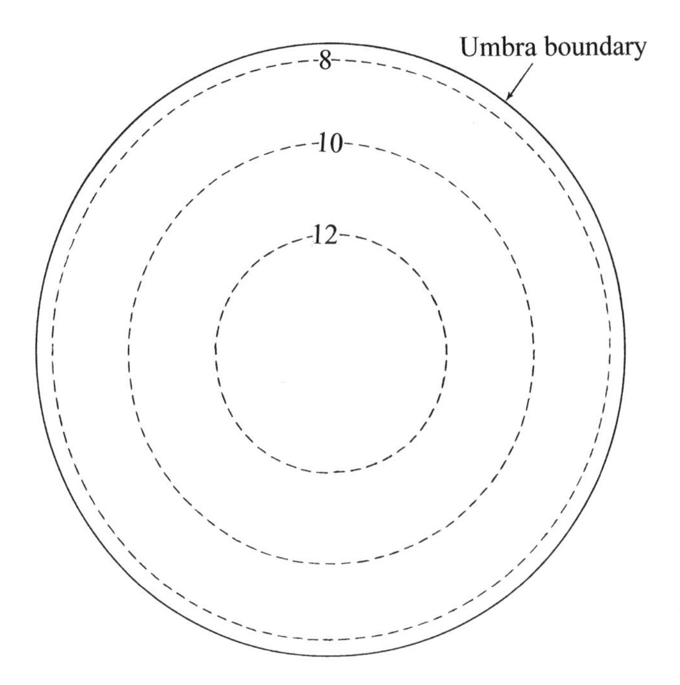

Fig. 5.13. Approximate brightness drop values (in stellar magnitudes) for different locations in the umbra. These are based on the results of Westfall (1982, 1988) and Schmude et al. (2000). All values are with respect to the uneclipsed Moon (Credit: Richard Schmude, Jr.).

March 2–3, 1942. Their results show the Moon dimmed by several stellar magnitudes just inside the umbra and dimmed by an additional amount near the umbra center. More recently, Westfall and this author used a photometer to measure the brightness of a small area of the Moon as it moved through the umbra. These studies show the umbra is darkest near the center. Approximate brightness drop values are shown in Fig. 5.13. Therefore, if the Moon moves through the umbra center, it would get darker than if it remains near the upper edge.

To sum up, our atmosphere affects the Moon's color and brightness during a total eclipse. The location of the Moon within the umbra also affects its color and brightness. Our atmosphere also affects the size and shape of the umbra. Therefore, **by studying the umbra and eclipses we may learn more about our atmosphere. Accurate measurements of the brightness and color of objects in our umbra should** yield tighter constraints on models of our atmosphere.

Measurement of Umbra Characteristics

Geostationary satellites serve as excellent probes of the umbra because they are just a few meters across. They subtend an angle of ~0.02 arc seconds. Therefore, astronomers may monitor the umbra at the ~0.1 km scale by measuring their brightness. The Moon, on the other hand, is an extended object. Thus, Moon measurements may yield umbra characteristics at the 20–1,000 km scale.

Two sources of sunlight fall into Earth's shadow. These are direct sunlight and light that our atmosphere either refracts or scatters. Direct, refracted and scattered all fall into the penumbra, whereas only refracted and scattered light fall into the

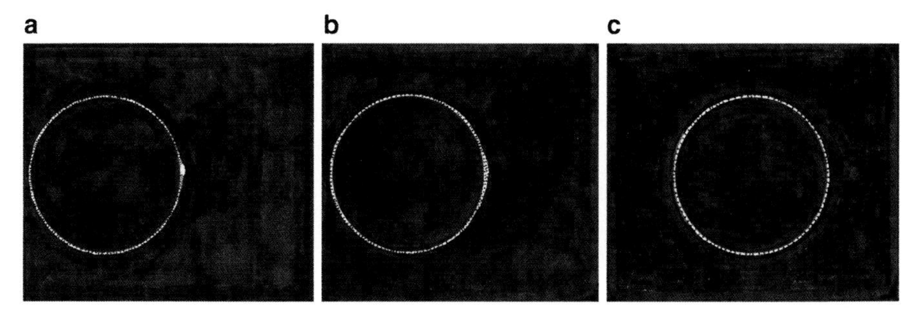

Fig. 5.14. Three views of Earth eclipsing the Sun. In drawing (**a**), Earth is blocking out part of the Sun on the right. The *faint ring* of light around our atmosphere is caused by refracted and scattered light. This is what an observer in the penumbra would see. In drawings (**b**) and (**c**), Earth completely blocks out the Sun. Refracted and scattered light is the cause of the faint ring of light (Credit: Richard Schmude, Jr.).

umbra. See Fig. 5.14. These drawings show how Earth, its atmosphere and our Sun would look in the penumbra and umbra. The solid white area in Fig. 5.14a is our Sun. About half of it is blocked. Stage A occurs in the penumbra. Stages B and C occur in our umbra. Figure 5.14b is the stage just after Earth blocks all direct sunlight. The amount of refracted and scattered light is greatest at the right edge. The dotted area around Earth represents refracted and scattered light. Based on measurements of our Moon and one geostationary satellite, the scattered and reflected light in Fig. 5.14b is about 2% of the light in Fig. 5.14a. This is also why a sharp drop in brightness does not occur as a lunar feature or satellite reaches U1. See Figs. 4.23 and 4.26.

How will the brightness of a geostationary satellite change as it enters Earth's shadow? To answer this question, several things should be considered. Recall from Chap. 4 that a satellite's brightness depends on its distance, phase and mode of reflection. The brightness of a geostationary satellite may rise to magnitude 7.5 just before entering Earth's shadow. Part of this is because it has nearly a full phase. A second factor to consider is the source of light entering Earth's shadow. If we assume a satellite reaches a stellar magnitude of 7.5 just before eclipse, it should drop to magnitude 11.5 at the darkest portion of the penumbra. Once inside the umbra, it should drop further depending on how deep it moves. Although the satellite will grow faint, it is still possible for an operator of a 0.3-m telescope with an appropriate camera to measure its brightness. These measurements will yield magnitude drops at different umbra depths. A third factor that may affect satellite brightness is the mode of reflection. If the amount of specular reflection increases during an eclipse this would affect satellite brightness. Finally, the amount of stratospheric ozone should affect brightness.

One should be aware of two sources of error in brightness measurements. One of these is changes in atmospheric transparency during an eclipse. This will cause a satellite to get brighter or dimmer and lead to error. Therefore, one should monitor the brightness of background stars during an eclipse. (Recall that Earth's shadow does not affect background stars.) If stars dim suddenly, this would indicate a change in atmospheric transparency. Therefore, the brightness of background stars may serve as a calibration source for atmospheric transparency. A second source of error is condensation on telescope parts. A dew shield will help prevent condensation on the objective lens or corrector plate.

Considering the points above, what may we learn about Earth's umbra? We will first examine some experimental results. Afterwards, we will explore research projects, grazing eclipses and brightness measurements of distant satellites.

Experimental Results

Using a 0.12 m refracting telescope this author was able to time eclipses of the geostationary satellite *Amazonas 2* by Earth. All measurements were made between October 1 and 16, 2011. Since *Amazonas 2* was bright, it was selected for study. It was also easy to identify since it had a fainter companion – *Amazonas 1.* See Fig. 5.15. Figure 5.16 shows a typical light curve of this satellite as it passed into Earth's shadow. The dimming started off very slow. As *Amazonas 2* moved deeper into the penumbra, it grew dimmer. During this time, refracted and scattered light was about 2% of the light reaching this satellite. When it approached U1, its brightness fluctuated by over a stellar magnitude. This fluctuation continued for about 15 s. These fluctuations are caused by changes in our atmosphere. They contain data. After the fluctuations the satellite faded below the limiting magnitude.

Figure 5.17 shows the general shape of Earth's shadow. The points represent the time interval between when *Amazonas 2* disappeared and when its right ascension was exactly opposite that of our Sun. A portion of the elliptical shadow is visible. Table 5.4 summarizes the writer's eclipse data. In some cases, *Amazonas 2* disappeared before it reached U1, and in other cases it disappeared a few seconds after reaching U1.

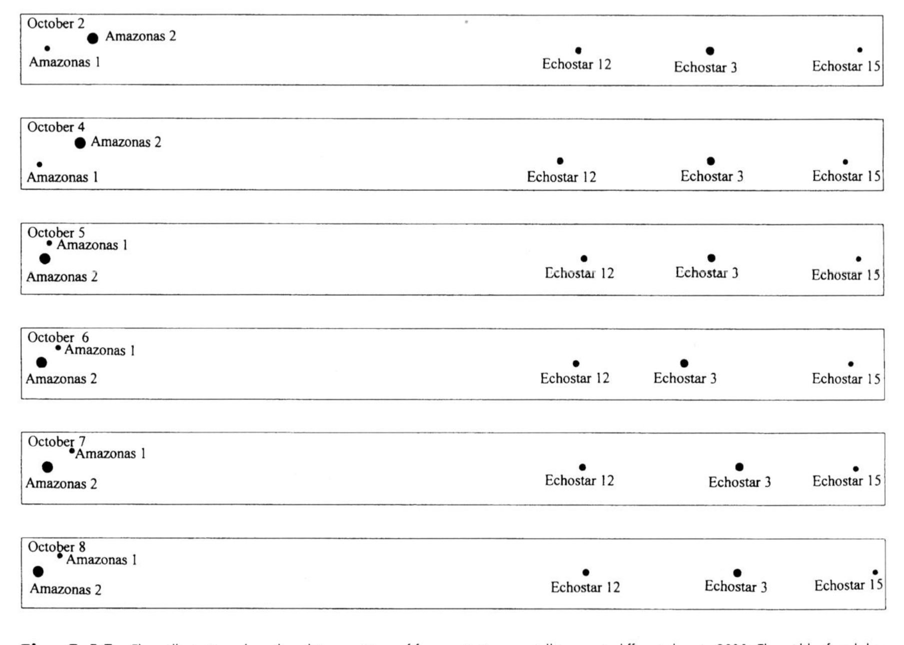

Fig. 5.15. These illustrations show the relative positions of five geostationary satellites on six different days in 2011. The width of each box equals 0.65° of longitude. The *size of the dot* represents the relative brightness (Credit: Richard Schmude, Jr.).

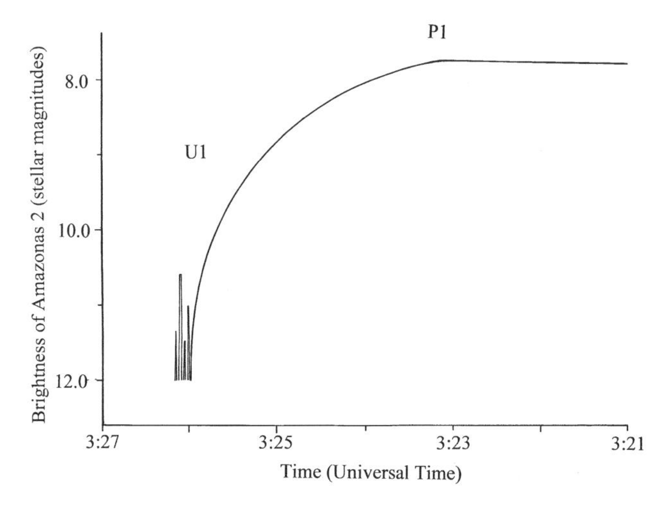

Fig. 5.16. An estimated light curve of *Amazonas 2* based on visual observations of an eclipse on October 7, 2011. The beginning of the penumbra phase (P1) took place at 3:23:12 UT, and the beginning of the umbra phase (U1) took place at 3:25:56 UT (Credit: Richard Schmude, Jr.).

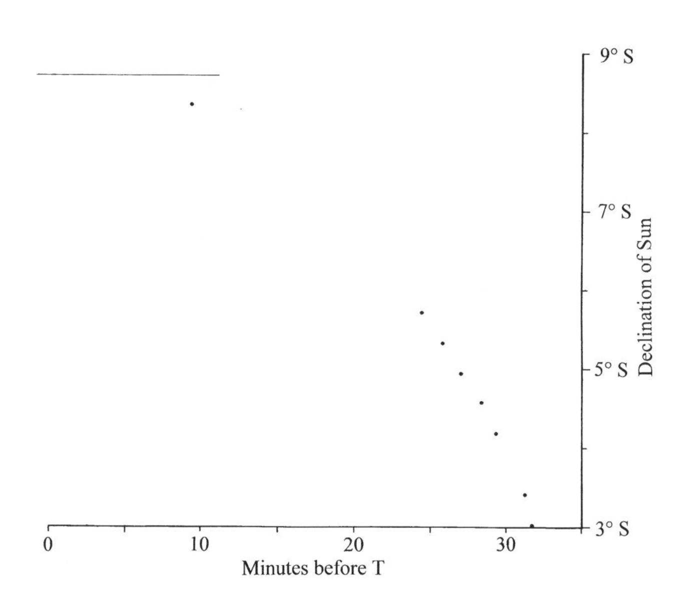

Fig. 5.17. This graph shows the number of minutes before T when *Amazonas 2* became invisible at different Sun declinations in October 2011. In this graph, T is the time when the satellite has exactly the opposite right ascension from our Sun and would be at mid-eclipse (Credit: Richard Schmude, Jr.).

Research Projects

There are three important research projects you can do using satellite eclipse data. One of these is to monitor the brightness of a satellite as it moves through the umbra. This may be done with the appropriate camera, software and telescope. Remember, a clock drive is not needed. Since satellites may grow quite dim, a large

Table 5.4. Predicted times of U1 for a solid airless Earth and observed time when *Amazonas 2* was no longer visible. All observations are in Universal Time. The faintest star seen in most observations was magnitude 12

Date (decimal date)	Predicted time of U1 (h:min:s)	Observed time of disappearance (h:min:s)
October 1.14, 2011	3:22:36	3:22:10
October 2.14, 2011	3:22:56	3:22:28
October 4.14, 2011	3:23:40	3:23:32
October 5.14, 2011	3:24:13	3:24:08
October 6.14, 2011	3:25:01	3:25:07
October 7.14, 2011	3:25:56	3:26:07
October 8.14, 2011	3:27:02	3:27:11
October 15.16, 2011	3:49:45	3:43:46

telescope aperture will give better results. The amount of dimming will depend on the characteristics of our atmosphere. With a sufficient amount of eclipse data, scientists may be able to construct better models of our atmosphere. For example, you can use eclipse measurements to determine how a large volcanic eruption or solar outburst affects our atmosphere. Unlike lunar eclipses, satellite eclipses occur almost 90 nights a year. Therefore, one can monitor day-to-day changes in the umbra. One can even monitor atmospheric characteristics over different areas of the world by observing eclipses at different times. Several hundred geostationary satellites orbit Earth. They cover all longitudes.

A second research project is to measure the satellite brightness with two or more color filters. This will yield color data. We know the Moon gets redder as it enters the umbra. Geostationary satellites should also grow redder as they move into the umbra. Color data will enable scientists to construct better models of our atmosphere.

One may also measure the central flash. This occurs when an object passes near the umbra's center. Astronomers have recorded central flashes of stars many times. In these cases, a planet with an atmosphere moves in front the star. According to Elliot (1979) when light is refracted around a planet's limb and arrives at the same location near the umbra's center, a central flash occurs. Geostationary satellites pass near the umbra's center within 1 day of equinox. Therefore, this is one more measurement that can be made. The magnitude and color of the central flash may give us information about the extinction of our atmosphere.

Grazing Eclipses

A grazing eclipse of a geostationary satellite is one that occurs when the Sun's declination is between 8.40° and 8.60°. When it equals 8.40°, the satellite just touches the boundary of the un-enlarged umbra. When the declination equals 8.60°, it touches the umbra boundary only when it is enlarged by 2.0%. Table 5.5 shows the dates and times when the Sun has a declination between 8.40° and 8.60° between 2013 and 2020. Grazing eclipses occur on four dates each year. Each opportunity lasts for about 13 h. For example, during the year 2016, grazing eclipses will occur on February 27, April 11, August 31 and October 14–15. A satellite may flicker for several seconds to perhaps a minute during a grazing eclipse. Occultations and flickering are described in *Uranus, Neptune and Pluto and How to Observe Them*

Table 5.5. Dates and times when the absolute value of the Sun's declination was between 8.40° and 8.60°. All times are in Universal Time

Year	Dates when 8.40° < ıδ8.60° >ı
2013	Feb. 26 (13:08) to Feb. 27 (1:56); Apr. 11 (11:20) to Apr. 12 (0:26); Aug. 31 (8:32) to Aug. 31 (21:49) and Oct. 14 (22:28) to Oct. 15 (11:26)
2014	Feb. 26 (19:19) to Feb. 27 (8:07); Apr. 11 (17:26) to Apr. 12 (6:33); Aug. 31 (14:32) to Sept. 1 (3:39) and Oct. 15 (4:38) to Oct. 15 (17:37)
2015	Feb. 27 (1:26) to Feb. 27 (14:14); Apr. 11 (23:37) to Apr. 12 (12:44); Aug. 31 (20:49) to Sept. 1 (10:06) and Oct. 15 (10:47) to Oct. 15 (23:45)
2016	Feb. 27 (7:33) to Feb. 27 (20:21); Apr. 11 (5:40) to Apr. 11 (18:46); Aug. 31 (3:03) to Aug. 31 (16:20) and Oct. 14 (16:59) to Oct. 15 (5:58)
2017	Feb. 26 (13:43) to Feb. 27 (2:30); Apr. 11 (11:54) to Apr. 12 (1:01); Aug. 31 (9:03) to Aug. 31 (22:20) and Oct. 14 (23:06) to Oct. 15 (12:04)
2018	Feb. 26 (19:50) to Feb. 27 (8:38); Apr. 11 (18:06) to Apr. 12 (7:12); Aug. 31 (15:16) to Sept. 1 (4:33) and Oct. 15 (5:14) to Oct. 15 (18:13)
2019	Feb. 27 (1:56) to Feb. 27 (14:44); Apr. 12 (0:01) to Apr. 12 (13:07); Aug. 31 (21:23) to Sept. 1 (10:40) and Oct. 15 (11:31) to Oct. 16 (0:29)
2020	Feb. 27 (8:02) to Feb. 27 (20:49); Apr. 11 (6:13) to Apr. 11 (19:20); Aug. 31 (3:29) to Aug. 31 (16:47) and Oct. 14 (17:37) to Oct. 15 (6:35)

 If flickering occurs, it would be an additional opportunity to learn more about our stratosphere and mesosphere.

On October 15, 2011, at 3:40–4:00 U.T., this author monitored a near-grazing eclipse of *Amazonas 2*. My limiting magnitude was ~12. The Sun's declination was 8.36°S. The satellite grew dimmer and flickered before disappearing. It was invisible for almost 13 min (3:43:46–3:56:40 U.T.). Therefore, even at the boundary of the umbra and penumbra, objects may dim by at least four stellar magnitudes. On the next night, the Sun's declination was 8.73°S. *Amazonas 2* dimmed about 1.5 stellar magnitudes, but it always remained visible. *Amazonas 1* was nearby. It was as bright as a star of magnitude 10.5. At times, it was near the limit of visibility. *Amazonas 2* remained visible the whole time because it did not enter the umbra.

Brightness Measurements of Distant Satellites

One research project that a few individuals have undertaken is to monitor the brightness of satellites and hazardous space debris. Unlike natural bodies, human-made objects often have a measurable brightness drop over a few years. Brightness changes are important from the standpoint of monitoring hazardous space debris.

Astronomers measure the brightness and other characteristics of distant satellites and report their results to the Minor Planet Center. This organization then publishes the results in the Distant Artificial Satellites Observation (DASO) circulars. These circulars are available online at http://www.minorplanetcenter.net/iau/DASO/DASO_OLD.html.

As mentioned in Chap. 4, the standard magnitude is the brightness a satellite has when it is 1.0 astronomical unit from our Sun, is 1,000 km from the observer and is at half phase. The standard V filter magnitude will be written as V(1, 1,000, 90°).

When specular reflection is insignificant, a satellite's brightness is expressed as

$$V(1,1000,90°) = V - 5 \times \mathrm{Log}\left(r \times \Delta / 0.0000066845\right) + \left(90° - \alpha\right) \times c_V \qquad (5.3)$$

In this equation, V is the measured brightness value in stellar magnitudes, r is the satellite-Sun distance, Δ is the satellite-observer distance, α is the solar phase angle and c_V is the solar phase angle coefficient for the V filter. Both r and Δ are in astronomical units (au). The 0.0000066845 term converts Δ from au to km. One should measure V and c_V for each satellite. The other terms are available in the ephemeris of specific satellites. One determines the c_V value by plotting the V(1, 1,000, α) values versus α. The V(1, 1,000, α) values are computed from

$$V(1,1000,\alpha) = V - 5 \times \mathrm{Log}\left(r \times \Delta / 0.0000066845\right) \qquad (5.4)$$

Figure 5.18 shows graphs of V(1, 1000, α) or R(1, 1000, α) versus α for *LCROSS*, *Epoxi* and *Spektr-R*. A linear equation is a satisfactory fit for all three satellites.

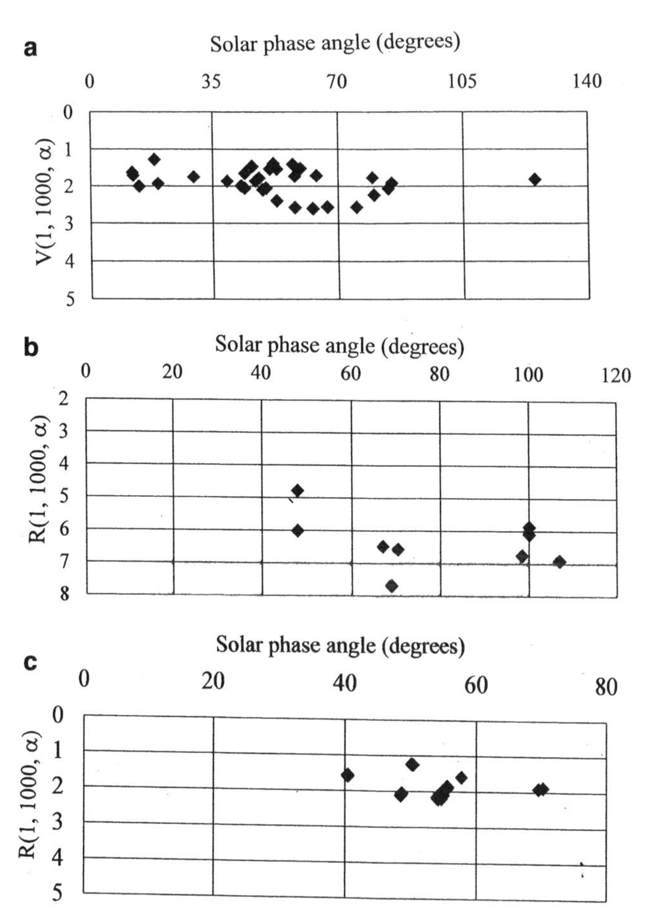

Fig. 5.18. The value of V(1, 1000, α) or R(1, 1,000, α) is plotted against the solar phase angle for the LCROSS (**a**), Epoxi/Deep Impact (**b**) and Spektr-R (**c**) satellites. The slope of the line corresponds to the solar phase angle coefficient in Eq. 5.3. These graphs are based on data in the DASO circulars along with distances and solar phase angles in the JPL Ephemeris Generator (Credit: DASO Circulars, JPL Ephemeris Generator and Richard Schmude, Jr.).

Table 5.6. Equations relating the standard magnitude to the solar phase angle for different satellites

Satellite	Equation	Source
LCROSS	$V(1, 1000, \alpha) = 1.67 + 0.014\alpha$	My analysis of data on the DASO circulars
LCROSS	$R(1, 1000, \alpha) = 0.74 + 0.0266\alpha$	My analysis of data on the DASO circulars
RK252A5	$R(1, 1000, \alpha) = 4.01 + 0.0106\alpha$	Miles (2011)[a]
Spektr-R	$R(1, 1000, \alpha) = 1.26 + 0.0113\alpha$	My analysis of data in the DASO circulars up to Dec. 6, 2011
Epoxi	$R(1, 1000, \alpha) = 5.42 + 0.0118\alpha$	My analysis of data in the DASO circulars

[a]The brightness is converted to a satellite standard magnitude

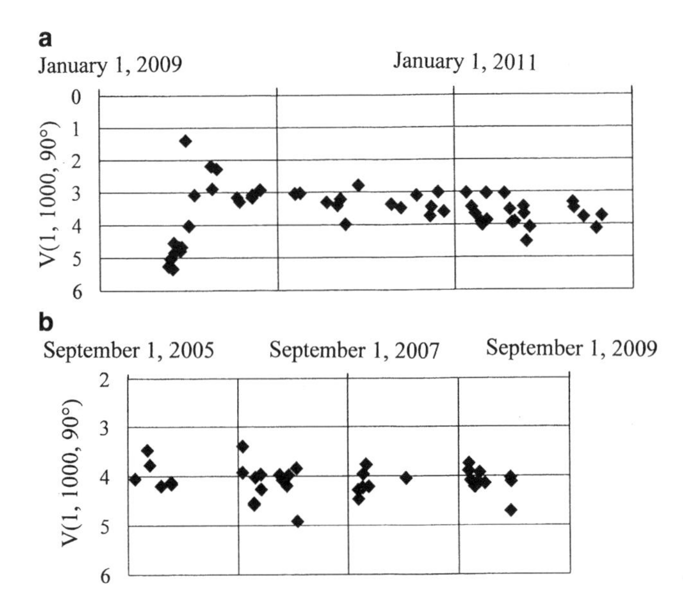

Fig. 5.19. The value of V(1, 1000, 90°) is plotted against the date for the Planck (**a**) and the Wilkinson Microwave Anisotropy Probe or WMAP (**b**) satellites. The normalized magnitudes of both satellites are computed using Eq. 5.3 and an assumed value of $c_V = 0.014$ magnitude/degree. Since the α values did not change much, this assumption has little influence on the final results (Credit: DASO Circulars, JPL Ephemeris Generator and Richard Schmude, Jr.).

The results are listed in Table 5.6. These equations are the first step in determining how the space environment affects satellite brightness over time.

Miles (2011) reports that the paint on RK252A5 (probably a spent rocket booster launched in the 1950s or 1960s.) grew darker by about a factor of three. This is equal to a brightness drop of 1.2 stellar magnitudes. This was motivation to look at other satellites. Figure 5.19 shows how the brightness of the *Planck* and the *Wilkinson Microwave Anisotropy Probe (WMAP)* satellites has changed over time. In both graphs, the normalized magnitude V(1, 1000, 90°) is plotted against the date.

Between October 9, 2009, and October 28, 2011, *Planck* darkened at a rate of 0.39 stellar magnitudes per year. *Planck* lifted off on May 14, 2009. Therefore, it dimmed substantially during its first 2.5 years in space.

The *WMAP* satellite grew darker at an average rate of 0.04 magnitudes/year. Unlike *Planck,* the results for *WMAP* started almost 4 years after this satellite was launched. Therefore, much of the darkening may have occurred before 2005. The results of *Planck*, *WMAP* and *RK252A5* (Miles 2011) suggest satellites darken

Table 5.7. Standard magnitudes of different satellites based on brightness measurements reported in the Distant Artificial Satellites Observation circulars. In this analysis, we assume solar phase angle coefficients of 0.014 and 0.011 magnitudes/degree for the V and R filters, respectively

	Satellite standard magnitude			
Satellite	V filter	R filter	Time interval since launch (years)	Source DASO circulars
LCROSS	2.9	3.1	<0.1	252, 255–6, 258–9, 262, 265–273, 288–9, 293, 295, 304, 208, 310–1, 314
Dawn	–	5.1	<0.1	117
Kepler	4.8	5.0	<0.1	201–203
Herschel	4.1	3.8	<0.1	222, 224, 227–30, 232–3, 236
Planck	4.9	4.6	<0.1	223, 227–8, 232, 236
Epoxi/Deep Impact	–	5.8	<0.1	19
Spektr-R	1.7	2.3	0.3	386, 390–2, 396–7
Stereo B	3.3	–	0.3	81–2, 87, 89
Genesis	6.2	5.1	3.1	6, 11
Rosetta	2.8	2.6	3.7	123–6, 128–9
Epoxi/Deep Impact	6.3	6.5	4	133, 135, 175–7, 179
WMAP	4.2	4.1	6.3	114, 116, 118, 120–1
NExT	–	4.5	9.9	184–8

over time. The darkening may be more rapid during the first 2 years in space and may taper off afterwards.

Both the solar phase angle and the time spent in space affect satellite brightness. Other factors such as shape, rotation and mode of reflection also affect brightness. More studies are needed to better understand the brightness of human-made objects in space.

Below you will find computed standard magnitudes of a few satellites. These are based on brightness values reported in the DASO circulars. The results are summarized in Table 5.7. *Spektr-R* is the brightest satellite in the table. This is because of its large diameter of 10 m.

Chapter 6

Computation of Eclipse Times

Geostationary satellites move into Earth's penumbra and umbra at specific times within 23 or 24 days of equinox. The deepest eclipses occur at equinox. The satellite path through Earth's shadow depends on the Sun's declination. It also depends on the exact satellite position. The observer's longitude and latitude, however, will not affect the eclipse time. In this chapter, we will explore how to compute eclipse times for a geostationary satellite.

Eclipse times are predicted to the nearest minute. More exact times require exact satellite positions. As mentioned in Chap. 4 and illustrated in Fig. 4.1, satellite positions may change a little. The website http://www.n2yo.com/satellites/?c=10 lists positions to the nearest 0.01°. One should plan to start watching for an eclipse a few minutes before the published eclipse times because of changing satellite positions.

We will first examine time zones, and afterwards, eclipses. Each time zone has a standard longitude. Standard longitudes are listed in Table 6.1. For the Eastern Time Zone, the standard longitude is 75°W. The W (west) is added because this zone is west of the Prime Meridian, which runs through Greenwich, England. The time depends on the location of the standard longitude. For example, all clocks in the Eastern Time Zone are 5 h behind those in Greenwich, England. Time no longer depends on the position of the Sun. Most of the United States, Canada and the United Kingdom recognize Daylight Savings Time. As of 2011, it begins on the second Sunday in March and ends on the first Sunday in November for the United States and Canada. In the United Kingdom, British Summer Time begins on the last Sunday in March and ends on the last Sunday in October. This time change should be included in predicted eclipse times.

The first step in computing eclipse times is to determine when the satellite's right ascension is exactly opposite that of our Sun (T). See Fig. 6.1. Segment LK in this drawing has a right ascension opposite to that of our Sun, and it bisects the segment between P1 and P4 and the segment between U1 and U4. Segment LK also crosses the center of Earth's shadow. We know the center of Earth's shadow transits the meridian twelve hours after the Sun. The appropriate equation for T is:

$$T = (E - 12\text{hours}) + 1\text{ hour} - 4.0\text{ minute / degree} \times (SL - st). \quad (6.1)$$

In this equation, E is the ephemeris transit time of the Sun, SL is the satellite longitude and st is the standard longitude. Both SL and st are in degrees. If SL and st

R. Schmude, Jr., *Artificial Satellites and How to Observe Them*, Astronomers' Observing Guides,
DOI 10.1007/978-1-4614-3915-8_6, © Springer Science+Business Media New York 2012

Table 6.1. Standard longitudes for a few different time zones

Time zone	Standard longitude (st)
Atlantic	60°W or −60°
Eastern	75°W or −75°
Central	90°W or −90°
Mountain	105°W or −105°
Pacific	120°W or −120°
Alaska (Anchorage)	135°W or −135°
Hawaii	150°W or −150°
United Kingdom	0° (Prime Meridian)
Western Europe (excluding the UK)	15°E or 15°
Belarus/Israel	30°E or 30°
India	82.5°E or 82.5°
China	120°E or 120°
Eastern Australia (Sydney)	150°E or 150°

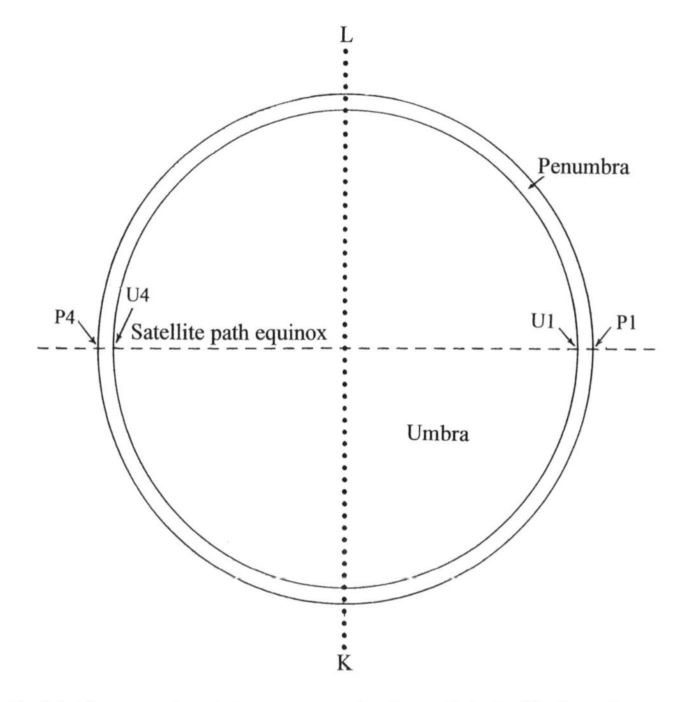

Fig. 6.1. The *dashed line* represents the path of a geostationary satellite. Segment LK, the *dotted line,* has a right ascension that is opposite to that of our Sun (Credit: Richard Schmude, Jr.).

are east of the Prime Meridian, their values would be positive; otherwise their values would be negative. The 12 h is added because we are interested in the time when Earth's shadow transits the meridian and 1 h is added because we want to convert to Daylight Savings Time, or British Summer Time. The quantity (E − 12 h) is called the equation of time. If one is interested in determining T for standard

Table 6.2. Declination and Ephemeris transit times of the Sun near equinox in March and September. The values for E are at the standard longitude and at standard time

Date	Sun declination	E (Ephemeris transit time of Sun)
20.0 days before March equinox	−7.812°	12 h 12 min 23 s
10.0 days before March equinox	−3.946°	12 h 10 min 06 s
March equinox	0.000°	12 h 07 min 17 s
10.0 days after March equinox	+3.924°	12 h 04 min 16 s
20.0 days after March equinox	+7.728°	12 h 01 min 25 s
20.0 days before September equinox	+7.616°	11 h 59 min 20 s
10.0 days before September equinox	+3.865°	11 h 55 min 53 s
September equinox	0.000°	11 h 52 min 21 s
10.0 days after September equinox	−3.888°	11 h 49 min 01 s
20.0 days after September equinox	−7.699°	11 h 46 min 13 s

time, the 1 h would not be added. The 4.0 min/degree is the average angular speed of the satellite with respect to the Earth-Sun line. The time (T) depends on the position of the Sun, Earth and satellite. It does not depend on where the observer is within a time zone.

Table 6.2 lists the declinations and Ephemeris transit times of the Sun for a few dates. One may use linear interpolation to compute transit times on other dates or consult the *Astronomical Almanac.*

Let's compute T for Echostar 15 for 10.0 days before the September equinox for the Eastern Time Zone.

$$T = 11\text{h } 55\text{m } 53\text{s} - 12 \text{ hours} + 1 \text{ hour} - 4.0 \text{ minute / degree} \times (-61.5° - -75.0°\text{W})$$

$$T = 11\text{h } 55\text{m } 53\text{s} - 12 \text{ hours} + 1 \text{ hour} - 4.0 \text{ minute / degree} \times (13.5°)$$

$$T = 11\text{h } 55\text{m } 53\text{s} - 12 \text{ hours} + 1 \text{ hour} - 54 \text{ minutes}$$

$$T = 0\text{h } 01\text{m } 53\text{s or } 12:01:53 \text{ A.M. Eastern Daylight Time}$$

The 11 h 55 min 53 s used above is the ephemeris transit time of the Sun exactly 10.0 days before the minute of equinox. One may compute T for other time zones; however, the correct standard longitude should be used.

Note especially that these values are in relation to the exact moment of equinox. For example, the equinox was at 9:05 UT on September 23, 2011. Therefore, 10 days before this equinox corresponds to 9:05 UT on September 13, 2011. The exact time of equinox shifts by a few hours each year.

The next task is to determine the time of P1, U1, U4 and P4. The ephemeris transit time of the Sun and the Earth-Sun distance changes throughout the year because of Earth's elliptical orbit. See Table 6.2. Figure 6.2 illustrates Earth's shadow and the paths a geostationary satellite will follow before and after equinox. The distance between P1 (or P4) and segment LK equals Pm and that between U1 (or U4) and segment LK equals Um. In order to determine when a satellite is at P1, you would subtract the time it takes for it to travel the distance Pm from T. Likewise, in order to compute the time of U1, you would subtract the time it takes for the satellite to travel the distance Um from T. Therefore, we will examine the steps needed to compute Pm and Um.

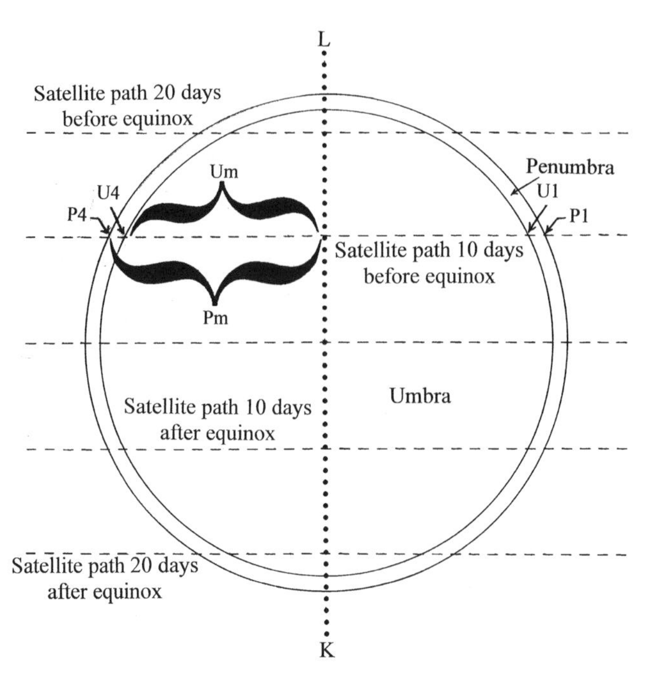

Fig. 6.2. The different paths a geostationary satellite follows at different times. Segment LK bisects Earth's shadow and has a right ascension that is opposite to that of our Sun (Credit: Richard Schmude, Jr.).

Earth's equatorial diameter is a little larger than its polar diameter. Therefore, an equation describing an ellipse is a better match for Earth's shadow than one describing a circle. (In this case, the Sun is assumed to have a perfectly spherical shape.) The equations for the elliptical shape of Earth's penumbra and umbra are:

$$\frac{X^2}{P_E^{\ 2}} + \frac{Y^2}{P_P^{\ 2}} = 1 \tag{6.2}$$

$$\frac{X^2}{U_E^{\ 2}} + \frac{Y^2}{U_P^{\ 2}} = 1 \tag{6.3}$$

In these equations, P_E and P_P are the equatorial and polar radii of Earth's penumbra, and U_E and U_P are the equatorial and polar radii of the umbra. The elliptical shapes of the penumbra and umbra are plotted on the X-Y plane in Fig. 6.2 and are centered on the origin. The Y axis is vertical and the X axis is horizontal.

The values of P_E, P_P, U_E and U_P are computed as:

$$P_E = r_E + \left\{ d \times \left[\left(r_S + r_E \right) / \, SE \right] \right\} \tag{6.4}$$

$$P_P = r_P + \left\{ d \times \left[\left(r_S + r_P \right) / \, SE \right] \right\} \tag{6.5}$$

$$U_E = r_E - \left\{ d \times \left[\left(r_S - r_E \right) / \, SE \right] \right\} \tag{6.6}$$

$$U_P = r_P - \left\{ d \times \left[\left(r_S - r_P \right) / \, SE \right] \right\} \tag{6.7}$$

In all cases, SE equals the Earth-Sun distance, d equals the radius of the satellite orbit, r_S is the radius of the Sun, r_E equals the equatorial radius of the Earth (6,378 km) and r_P equals the polar radius of Earth (6,357 km). Values of P_E, P_P, U_E and U_P are listed for several dates in Table 7.3.

The lengths of Pm and Um should be determined. One of the dashed lines in Fig. 6.2 shows the path a geostationary satellite follows 10.0 days before the September equinox. In order to compute Pm, one should determine the intersection of the line running through P1 and P4 and the ellipse. The equation of the line running through P1 and P4 is:

$$Y = d \times \sin(\delta) \tag{6.8}$$

In this equation, d is the radius of the satellite's orbit and δ is the Sun's declination. To compute the X coordinate of P1, one must substitute Eq. 6.8 into 6.2:

$$\frac{X^2}{P_E^{\ 2}} + \frac{\left(d \times \sin(\delta)\right)^2}{P_P^{\ 2}} = 1 \tag{6.9}$$

or rearrange it as:

$$X = \left\{P_E^{\ 2} \times \left[1 - \left\{\left(d \times \sin(\delta)\right)^2 / P_P^{\ 2}\right\}\right]\right\}^{0.5} \tag{6.10}$$

The value of X is the X coordinate of P1 and it equals Pm. The distance between P4 or P1 and segment LK is Pm. In a similar way, one may find the X component of U1 and determine Um by substituting Eq. 6.8 into 6.3. This equals Um. Therefore, you can write:

$$Pm = \left\{P_E^{\ 2} \times \left[1 - \left\{\left(d \times \sin(\delta)\right)^2 / P_P^{\ 2}\right\}\right]\right\}^{0.5} \tag{6.11}$$

$$Um = \left\{U_E^{\ 2} \times \left[1 - \left\{\left(d \times \sin(\delta)\right)^2 / U_P^{\ 2}\right\}\right]\right\}^{0.5} \tag{6.12}$$

Once the values of Pm and Um are determined you can compute the times of P1, U1, U4 and P4 as:

$$\text{Time of P1} = T - \left(Pm \div 3.066\ \text{km/s}\right) \tag{6.13}$$

$$\text{Time of P4} = T + \left(Pm \div 3.066\ \text{km/s}\right) \tag{6.14}$$

$$\text{Time of U1} = T - \left(Um \div 3.066\ \text{km/s}\right) \tag{6.15}$$

$$\text{Time of U4} = T + \left(Um \div 3.066\ \text{km/s}\right) \tag{6.16}$$

In all four equations, the 3.066 km/s is the average speed of a geostationary satellite with respect to Earth's shadow.

Values of Pm and Um for several times near the March and September equinoxes are listed in Table 6.3.

The method for computing the time of P1, P4, U1 and U4 are summarized in Fig. 6.3.

Table 6.3. Critical characteristics of Earth's penumbra and umbra, which are needed to predict eclipses of geostationary satellites

Date	Pm (km)	Um (km)	d × sin(δ) (km)	Sun-Earth distance (km)	Penumbra P_E (km)	Penumbra P_P (km)	Umbra U_E (km)	Umbra U_P (km)
20 days before March equinox	3,196	2,269	5,731	1.4821×10^8	6,578	6,557	6,182	6,161
10 days before March equinox	5,898	5,453	2,902	1.4858×10^8	6,577	6,556	6,182	6,161
March equinox	6,577	6,183	0	1.4898×10^8	6,577	6,556	6,183	6,162
10 days after March equinox	5,905	5,463	2,885	1.4942×10^8	6,576	6,555	6,183	6,162
20 days after March equinox	3,300	2,424	5,670	1.4984×10^8	6,576	6,555	6,184	6,163
20 days before September equinox	3,434	2,610	5,588	1.5092×10^8	6,574	6,553	6,185	6,164
10 days before September equinox	5,925	5,488	2,842	1.5053×10^8	6,575	6,554	6,185	6,164
September equinox	6,576	6,184	0	1.5013×10^8	6,575	6,554	6,184	6,163
10 days after September equinox	5,918	5,478	2,859	1.4971×10^8	6,576	6,555	6,184	6,163
20 days after September equinox	3,336	2,471	5,649	1.4927×10^8	6,576	6,555	6,183	6,162

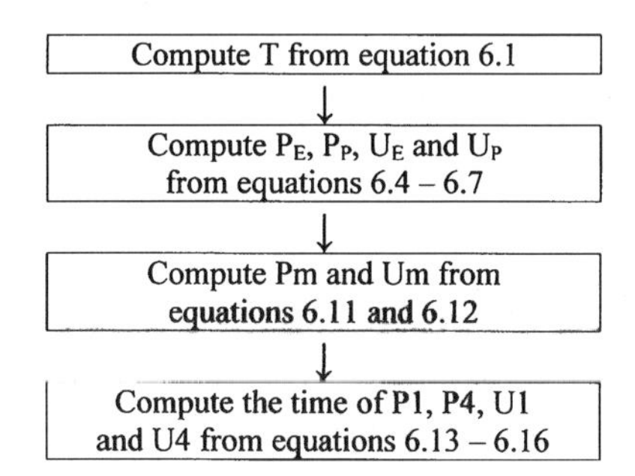

Fig. 6.3. Procedure used in computing the eclipse times of geostationary satellites (Credit: Richard Schmude, Jr.).

Example Calculation

Let us work out an example showing how P1, P4, U1 and U4 are computed for *Echostar 15,* 10.0 days before the September equinox.

One first computes T from Eq. 6.1 (first box in Fig. 6.3):

$$T = (E - 12\text{ hours}) + 1\text{ hour} - 4.0\text{ minute/degree} \times (SL - st)$$

$$T = 11\text{h } 55\text{m } 53\text{s} - 12\text{ hours} + 1\text{ hour} - 4.0\text{ minute/degree} \times (-61.5° - -75.0°\text{W})$$

$$T = 0\text{h } 01\text{m } 53\text{s} \text{ or } 12:01:53 \text{ A.M. Eastern Daylight Time } \textit{as illustrated earlier.}$$

The values of P_E, P_P, U_E and U_P are computed from Eqs. 6.4, 6.5, 6.6, and 6.7 (second box in Fig. 6.3) as:

$$P_E = 6378\text{ km} + \{42{,}164\text{ km} \times [(696{,}265\text{ km} + 6378\text{ km})/(1.5053 \times 10^8\text{ km})]\}$$

$$P_E = 6575\text{ km}$$

$$P_P = 6357\text{ km} + \{42{,}164\text{ km} \times [(696{,}265\text{ km} + 6357\text{ km})/(1.5053 \times 10^8\text{ km})]\}$$

$$P_P = 6554\text{ km}$$

$$U_E = 6378\text{ km} - \{42{,}164\text{ km} \times [(696{,}265\text{ km} - 6378\text{ km})/(1.5053 \times 10^8\text{ km})]\}$$

$$U_E = 6185\text{ km}$$

$$U_P = 6357\text{ km} - \{42{,}164\text{ km} \times [(696{,}265\text{ km} - 6357\text{ km})/(1.5053 \times 10^8\text{ km})]\}$$

$$U_P = 6164\text{ km}$$

The third box in Fig. 6.3 calls for the evaluation of Pm and Um. The value of Pm is computed from Eq. 6.11 as:

$$Pm = \{(6575\text{ km})^2 \times [1 - \{(42{,}164\text{ km} \times \sin(3.865°))^2/(6554\text{ km})^2\}]\}^{0.5}$$

$$Pm = \{(6575\text{ km})^2 \times [1 - \{(2842\text{ km})^2/(6554\text{ km})^2\}]\}^{0.5}$$

$$Pm = \{(6575\text{ km})^2 \times [1 - \{0.18805\}]\}^{0.5}$$

$$Pm = \{35{,}101{,}250\text{ km}^2\}^{0.5} = 5925\text{ km}$$

The value of Um is computed from Eq. 6.12 as:

$$\text{Um} = \left\{(6185\ \text{km})^2 \times \left[1 - \left\{(42{,}164\ \text{km} \times \sin(3.865°))^2 / (6164\ \text{km})^2\right\}\right]\right\}^{0.5}$$

$$\text{Um} = \left\{(6185\ \text{km})^2 \times \left[1 - \left\{(2842\ \text{km})^2 / (6164\ \text{km})^2\right\}\right]\right\}^{0.5}$$

$$\text{Um} = \left\{(6185\ \text{km})^2 \times [1 - \{0.2126\}]\right\}^{0.5}$$

$$\text{Um} = \left\{30{,}121{,}564\ \text{km}^2\right\}^{0.5} = 5488\ \text{km}$$

Finally, the fourth box in Fig. 6.3 calls for the evaluation of the times of P1, P4, U1 and U4. The Time of P1 is computed from Eq. 6.13 as:

$$\text{Time of P1} = \text{T} - (5925 / 3.066\ \text{km/s})$$
$$\text{Time of P1} = 0:01:53 - 1932\text{s}$$
$$\text{Time of P1} = 23:29:41.$$

You can compute the Time of P4 from Eq. 6.14 as:

$$\text{Time of P4} = \text{T} + (5925 / 3.066\ \text{km/s})$$
$$\text{Time of P4} = 0:01:53 + 1932\text{s}$$
$$\text{Time of P4} = 0:34:05.$$

You can compute the Time of U1 from Eq. 6.15 as:

$$\text{Time of U1} = \text{T} - (5488 / 3.066\ \text{km/s})$$
$$\text{Time of U1} = 0:01:53 - 1790\text{s}$$
$$\text{Time of U1} = 23:32:03.$$

You can compute the Time of U4 from Eq. 6.16 as:

$$\text{Time of U4} = \text{T} + (5488 / 3.066\ \text{km/s})$$
$$\text{Time of U4} = 0:01:53 + 1790\text{s}$$
$$\text{Time of U4} = 0:31:43.$$

References

Abe S, Mukai T, Hirata N et al (2006a) Mass and local topography measurements of Itokawa by Hayabusa. Science 312:1344–1347
Abe M, Takagi Y, Kitazato K et al (2006b) Near-infrared spectral results of asteroid Itokawa from the Hayabusa spacecraft. Science 312:1334–1338
Angelo JA Jr (2000) Encyclopedia of space exploration. Facts on File, New York
Angelo JA Jr (2006) Encyclopedia of space and astronomy. Infobase, New York
Ashbrook J (1964) Measuring the Earth's shadow. Sky Telesc 27(3):156–160
(2008) Astronomical almanac for the year 2010. U. S. Govt. Printing Office, Washington, DC
Atkins P, de Paula J (2002) Physical chemistry, 7th edn. W H Freeman and Company, New York
Baker D (1981) The history of manned space flight. Crown, New York
Barabash S, Bhardwaj A, Wieser M et al (2009) Investigation of the solar wind – moon interaction onboard Chandrayaan-1 mission with the SARA experiment. Curr Sci 96:526–532
Barbier D (1961) Photometry of lunar eclipses. In: Kuiper GP, Middlehurst BM (eds) Planets and satellites. The University of Chicago Press, Chicago, pp 249–271
Barnes JW, Brown RH, Soderblom JM et al (2009) Shoreline features of Titan's Ontario Lacus from Cassini/VIMS observations. Icarus 201:217–225
Barnes JW, Brown RH, Soderblom L et al (2008) Spectroscopy morphometry and photoclinometry of Titan's dunefields from Cassini VIMS. Icarus 195:400–414
Beatty JK (2001) NEAR falls for Eros. Sky Telesc 101(5):34–37
Beatty JK (2005) Hayabusa arrives at asteroid Itokawa. Sky Telesc 110(6):17
Beatty JK (2006a) Hayabusa mission gets long delay. Sky Telesc 111(3):24
Beatty JK (2006b) SMART 1's grand finale. Sky Telesc 112(6):20
Beatty JK (2007) Celebrating Cassini. Sky Telesc 114(6):30–33
Beatty JK (2008) A Martian wonderland. Sky Telesc 116(4):22–24
Beatty JK (2010a) NASA slams the moon. Sky Telesc 119(2):28–32
Beatty JK (2010b) Mars under the microscope. Sky Telesc 120(3):20–25
Beatty JK, Chaikin A (1990) The new solar system, 3rd edn. Cambridge University Press, Cambridge, UK
Bhandari N (2005) Chandrayaan-1: science goals. J Earth Syst Sci 114:699–709
Bhardwaj A, Barabash S, Futaana Y et al (2005) Low energy neutral atom imaging on the Moon with the SARA instrument aboard Chandrayaan-1 mission. J Earth Syst Sci 114:749–760
Blades JC (2008) Fixing Hubble one last time. Sky Telesc 116(4):26–31
Brown RH, Baines KH, Bellucci G et al (2005) The Cassini visual and infrared mapping spectrometer (VIMS) investigation. Space Sci Rev 115:111–168
Brown RH, Soderblom LA, Soderblom JM et al (2008) The identification of liquid ethane in Titan's Ontario Lacus. Nature 454:607–610
Brown TL, LeMay HE Jr, Bursten BE et al (2006) Chemsitry the central science, 10th edn. Pearson Prentice Hall, Upper Saddle River
Bullard FM (1976) Volcanoes of the Earth. University of Texas Press, Austin, p 518
Burchell MJ, Robin-Williams R, Foing BH (2010) The SMART-1 lunar impact. Icarus 207:28–38
Clark BE, Grant KB (2005) Japan's asteroid archaeologist. Sky Telesc 109(6):34–37
Clery D (2011) Russia launches a telescope, decades in the making. Science 333:512

R. Schmude, Jr., *Artificial Satellites and How to Observe Them*, Astronomers' Observing Guides,
DOI 10.1007/978-1-4614-3915-8, © Springer Science+Business Media New York 2012

Cooper T (2004) The two total lunar eclipses of 2003. Monthly Notes of the Astronomical Society of Southern Africa 63(1, 2):8–11. See the website: http://www.netspeed.com.au/minnah/2003/pdf%20files/Cooper1.pdf
Cooper T, Geyser M (2004) Size and shape of the Umbra during a lunar eclipse. See the website: http://www.netspeed.com.au/minnah/2003/pdf%20files/Cooper2.pdf
Corliss WR (1967) Nuclear propulsion for space. U. S. Atomic Energy Commission, Oak Ridge
Cutnell JD, Johnson KW (2007) Physics, 7th edn. Wiley, Hoboken
Dachev T, Tomov B, Dimitrov P et al (2009) Monitoring lunar radiation environment: RADOM instrument on Chandrayaan-1. Curr Sci 96:544–546
Dambeck T (2008) Gaia's mission to the Milky Way. Sky Telesc 115(3):36–39
Damon TD (1989) Introduction to space: the science of spaceflight. Orbit Book, Malabar
David L (2008) China's space leadership. In: Mari C (ed) The next space age. H. W. Wilson, New York, pp 35–37
de Carvalho H (2003) Brightness of the total lunar eclipse of May 15–16, 2003. Online report at http://www.geocities.ws/lunissolar2003/Brightness_of_the_Total_Lunar_Eclipse_of_May_15_16_2003.pdf
Del Río-Gaztelurrutia T, Legarreta J, Hueso R et al (2010) A long-lived cyclone in Saturn's atmosphere: observations and models. Icarus 209:665–681
Demura H, Kobayashi S, Nemoto E et al (2006) Pole and global shape of 25143 Itokawa. Science 312:1347–1349
di Cicco D (1989) August's eclipsed moon. Sky Telesc 78(5):548–550
Dobbins TA, Parker DC, Capen CF (1988) Introduction to observing and photographing the solar system. Willmann-Bell, Richmond
Dougherty MK, Achilleos N, Andre N (2005) Cassini magnetometer observations during Saturn orbit insertion. Science 307:1266–1270
Dougherty MK, Kellock S, Southwood DJ et al (2004) The Cassini magnetic field investigation. Space Sci Rev 114:331–383
Elachi C, Allison MD, Borgarelli L et al (2005) Radar: the Cassini Titan radar mapper. Space Sci Rev 115:71–110
Elliot JL (1979) Stellar occultation studies of the solar system. Annu Rev Astron Astrophys 17:445–475
Esposito LW, Barth CA, Colwell JE et al (2005a) The Cassini ultraviolet imaging spectrograph investigation. Space Sci Rev 115:299–361
Esposito LW, Colwell JE, Larsen K et al (2005b) Ultraviolet imaging spectroscopy shows an active saturnian system. Science 307:1251–1255
Farrell SL, Laxon SW, McAdoo DC et al (2009) Five years of arctic sea ice freeboard measurements from the ice, cloud and land elevation satellite. J Geophys Res 114:C04008. doi:10.1029/2008JC005074
Fienberg RT (2005) Hubble gets a shot in the arm. Sky Telesc 109(3):18
Fischer D, Duerbeck H (1996) Hubble: a new window to the universe. Springer, New York
Fix JD (2011) Astronomy: journey to the cosmic frontier, 6th edn. McGraw Hill, New York
Flanders T (2005) Lunar eclipse science. Sky Telesc 109(3):84–85
Flasar FM, Achterberg RK, Conrath BJ et al (2005a) Temperatures, winds, and composition in the saturnian system. Science 307:1247–1251
Flasar FM, Kunde VG, Abbas MM et al (2005b) Exploring the Saturn system in the thermal infrared: the composite infrared spectrometer. Space Sci Rev 115:169–297
Foing BH, Racca GD, Marini A et al (2005) SMART-1 after lunar capture: first results and perspectives. J Earth Syst Sci 114:689–697
Fujiwara A, Kawaguchi J, Yeomans DK et al (2006) The rubble-pile asteroid Itokawa as observed by Hayabusa. Science 312:1330–1334
Galimov EM (2005) Luna-Glob project in the context of the past and present lunar exploration in Russia. J Earth Syst Sci 114:801–806
Gallagher JS (1992) Diffuse luminous objects having angular velocities similar to meteors. J Assoc Lunar Planet Obs 36:115–116

Gardner JP (2010) Finding the first galaxies. Sky Telesc 119(1):24–30
Gefter A (2005) Putting Einstein to the test. Sky Telesc 110(1):32–40
Gombosi TI, Hansen KC (2005) Saturn's variable magnetosphere. Science 307:1224–1226
Goody RM, Walker JCG (1972) Atmospheres. Prentice-Hall, Englewood Cliffs
Goswami JN, Annadurai M (2009) Chandrayaan-1: India's first planetary science mission to the moon. Curr Sci 96:486–491
Goswami JN, Banerjee D, Bhandari N et al (2005) High energy X-ray spectrometer on the Chandrayaan-1 mission to the moon. J Earth Syst Sci 114:733–738
Gove P - Editor in Chief (1971) Webster's third new international dictionary of the English language, G & C Merriam Co., Springfield MA
Graham FG (1987) Two lunar eclipses: 1983 JUN 25 and 1986 APR 24. J Assoc Lunar Planet Obs 32(1–2):28–30
Graham FG (1995) Analysis of the August 17, 1989 total lunar eclipse. J Assoc Lunar Planet Obs 38(2):61–64
Grande M, Maddison BJ, Sreekumar P et al (2009) The Chandrayaan-1 x-ray spectrometer. Curr Sci 96:517–519
Grifantini K (2007) Destination: Mars. Sky Telesc 114(5):30–31
Grün E (2007) Solar system dust. In: McFadden LA, Weissman PR, Johnson TV (eds) Encyclopedia of the solar system, 2nd edn. Elsevier, Amsterdam, pp 621–636
Guerlet S, Fouchet T, Bézard B et al (2009) Vertical and meridional distribution of ethane, acetylene and propane in Saturn's stratosphere from CIRS /Cassini limb observations. Icarus 203:214–232
Guerlet S, Fouchet T, Bézard B et al (2010) Meridional distribution of CH_3C_2H and C_4H_2 in Saturn's stratosphere from CIRS/Cassini limb and nadir observations. Icarus 209:682–695
Gurnett DA, Kurth WS, Hospodarksy GB et al (2005) Radio and plasma wave observations at Saturn from Cassini's approach and first orbit. Science 307:1255–1259
Gurnett DA, Kurth WS, Kirchner DL et al (2004) The Cassini radio and plasma pave investigation. Space Sci Rev 114:395–463
Haas WH (1982) Two reports from England on the January 9, 1982 total lunar eclipse. J Assoc Lunar Planet Obs 29(5–6):128–129
Haigh JD, Winning AR, Toumi R et al (2010) An influence of solar spectral variations on radiative forcing of climate. Nature 467:696–699
Hanlon M (2001) The worlds of *Ga*lileo. St. Martin's Press, New York
Harrje DT (2001) Rocket, vol 23, Encyclopedia Americana. Grolier, Danbury, pp 602–612
Hewitt PG, Lyons S, Suchocki J et al (2007) Conceptual integrated science. Addison Wesley, San Francisco
Hinshaw G, Naeye R (2008) Decoding the oldest light in the universe. Sky Telesc 115(5):18–23
Huixian S, Shuwu D, Jianfeng Y et al (2005) Scientific objectives and payloads of Chang'E-1 lunar satellite. J Earth Syst Sci 114:787–794
James CR (2007) Solar forecast: storm ahead. Sky Telesc 114(1):24–30
James N, Mason J (2003) Light levels at the 2002 December 4 eclipse. J Br Astron Assoc 113:195
Janssen S, Liu ML, Ross S et al (eds) (2011) The world Almanac and book of facts 2011. World Almanac Books®, New York
Johnston LR (2005) GALEX sees ultraviolet explosion. Sky Telesc 110(3):17
Johnston LR (2006) Swift: the satellite that's always on call. Sky Telesc 111(1):48–50
Kamalakar JA, Bhaskar KVS, Prasad ASL et al (2005) Lunar ranging instrument for Chandrayaan-1. J Earth Syst Sci 114:725–731
Kamalakar JA, Prasad ASL, Bhaskar KVS et al (2009) Lunar laser ranging instrument (LLRI): a tool for the study of topography and gravitational field of the moon. Curr Sci 96:512–516
Karkoschka E, Aguirre E (1996) Earth's swollen shadow. Sky Telesc 92(3):98–100
Kelly P (2007) Observer's handbook 2008. The Royal Astronomical Society of Canada, Toronto
Kempf S, Srama R, Postberg F et al (2005) Composition of saturnian stream particles. Science 307:1274–1276

Kennedy K (2010) Noctilucent cloud over Britain and Western Europe, 2006–2008. J Br Astron Assoc 120:152–156
Kerrod R (2004) Hubble: the mirror on the universe. Firefly Books, Buffalo
Kliore AJ, Anderson JD, Armstrong JW et al (2005) Cassini radio science. Space Sci Rev 115:1–70
Krishna A, Gopinath NS, Hegde NS et al (2005) Imaging and power generation strategies for Chandrayaan-1. J Earth Syst Sci 114:739–748
Koutchmy S, Nikol'skij GM (1983) The night sky from Salyut. Sky Telesc 65:23–25
Krimigis SM, Mitchell DG, Hamilton DC et al (2004) Magnetosphere imaging instrument (MIMI) on the Cassini mission to Saturn/Titan. Space Sci Rev 114:233–329
Kumar ASK, Chowdhury AR (2005a) Terrain mapping camera for Chandrayaan-1. J Earth Syst Sci 114:717–720
Kumar ASK, Chowdhury AR (2005b) Hyper-spectral imager in visible and near-infrared band for lunar compositional mapping. J Earth Syst Sci 114:721–724
Kumar ASK, Chowdhury AR, Banerjee A et al (2009a) Terrain mapping camera: a stereoscopic high-resolution instrument on Chandrayaan-1. Curr Sci 96:492–495
Kumar ASK, Chowdhury AR, Banerjee A et al (2009b) Hyper spectral imager for lunar mineral mapping in visible and near infrared band. Curr Sci 96:496–499
Kumar Y, MIP Project Team (2009) The moon impact probe on Chandrayaan-1. Curr Sci 96:540–543
Lewis RS (1990) Space in the 21st century. Columbia University Press, New York
Lide DR (ed) (2008) CRS handbook of chemistry and physics, 89th edn. CRS Press, Boca Raton, Editor in Chief
Lorenz RD, Newman C, Lunine JI (2010) Threshold of wave generation on Titan's lakes and seas: effect of viscosity and implications for Cassini observations. Icarus 207:932–937
Lutskij V (1982) Astronomy with Salyut 6. Sky Telesc 63:33–34
MacRobert A (2007a) March's twilight lunar eclipse. Sky Telesc 113(3):60–61
MacRobert A (2007b) A lunar eclipse in the dawn. Sky Telesc 114(2):50–53
Mall U, Banaszkiewic M, Bronstad K et al (2009) Near infrared spectrometer SIR-2 on Chandrayaan-1. Curr Sci 96:506–511
Mallama A, Caprette D, Sinnott R (1993) Anatomy of a lunar eclipse. Sky Telesc 86(3):76–78
McDowell J (2004) Mission update. Sky Telesc 108(6):26
McDowell J (2005a) Mission update. Sky Telesc 109(1):28
McDowell J (2005b) Mission update. Sky Telesc 109(2):28
McDowell J (2005c) Mission update. Sky Telesc 109(3):27
McDowell J (2005d) Mission update. Sky Telesc 109(4):29
McDowell J (2005e) Mission update. Sky Telesc 109(5):22
McDowell J (2005f) Mission update. Sky Telesc 109(6):26
McDowell J (2005g) Mission update. Sky Telesc 110(2):25
McDowell J (2005h) Mission update. Sky Telesc 110(3):24
McDowell J (2005i) Mission update. Sky Telesc 110(4):24
McDowell J (2005j) Mission update. Sky Telesc 110(5):26
McDowell J (2005k) Mission update. Sky Telesc 110(6):26
McDowell J (2006a) Mission update. Sky Telesc 111(2):24
McDowell J (2006b) Mission update. Sky Telesc 111(3):26
McDowell J (2006c) Mission update. Sky Telesc 111(4):24
McDowell J (2006d) Mission update. Sky Telesc 111(5):26
McDowell J (2006e) Mission update. Sky Telesc 112(1):26
McDowell J (2006f) Mission update. Sky Telesc 112(2):24
McDowell J (2006g) Mission update. Sky Telesc 112(3):24
McDowell J (2006h) Mission update. Sky Telesc 112(5):22
McDowell J (2006i) Mission update. Sky Telesc 112(6):30
McDowell J (2007a) Mission update. Sky Telesc 113(1):28
McDowell J (2007b) Mission update. Sky Telesc 113(2):26
McDowell J (2007c) Mission update. Sky Telesc 113(3):24

McDowell J (2007d) Mission update. Sky Telesc 113(4):24
McDowell J (2007e) Mission update. Sky Telesc 114(1):22
McDowell J (2007f) Mission update. Sky Telesc 114(2):22
McDowell J (2007g) Mission update. Sky Telesc 114(3):20
McDowell J (2007h) Mission update. Sky Telesc 114(4):17
McDowell J (2007i) Mission update. Sky Telesc 114(5):17
McDowell J (2007j) Mission update. Sky Telesc 114(6):17
McDowell J (2008a) Mission update. Sky Telesc 115(2):19
McDowell J (2008b) Mission update. Sky Telesc 115(3):21
McDowell J (2008c) Mission update. Sky Telesc 115(4):17
McDowell J (2008d) Mission update. Sky Telesc 115(5):15
McDowell J (2008e) Mission update. Sky Telesc 115(6):15
McDowell J (2008f) Mission update. Sky Telesc 116(1):15
McDowell J (2008g) Mission update. Sky Telesc 116(2):15
McDowell J (2008h) Mission update. Sky Telesc 116(3):15
McDowell J (2008i) Mission update. Sky Telesc 116(4):18
McDowell J (2008j) Mission update. Sky Telesc 116(5):22
McDowell J (2008k) Mission update. Sky Telesc 116(6):18
McDowell J (2009a) Mission update. Sky Telesc 117(1):18
McDowell J (2009b) Mission update. Sky Telesc 117(2):15
McDowell J (2009c) Mission update. Sky Telesc 117(3):18
McDowell J (2009d) Mission update. Sky Telesc 117(4):17
McDowell J (2009e) Mission update. Sky Telesc 117(5):20
McDowell J (2009f) Mission update. Sky Telesc 117(6):16
McDowell J (2009g) Mission update. Sky Telesc 118(1):18
McDowell J (2009h) Mission update. Sky Telesc 118(3):16
McDowell J (2009i) Mission update. Sky Telesc 118(5):16
McDowell J (2009j) Mission update. Sky Telesc 118(6):18
McDowell J (2010a) Mission update. Sky Telesc 119(1):20
McDowell J (2010b) Mission update. Sky Telesc 119(2):16
McFadden LA, Weissman PR, Johnson TV (2007) Encyclopedia of the solar system, 2nd edn. Elsevier, Amsterdam
Miles R (2011) The unusual case of 'asteroid' 2010 KQ: a newly-discovered artificial object orbiting our Sun. J Br Astron Assoc 121:350–354
Miller JK, Konopliv AS, Antreasian PG et al (2002) Determination of shape, gravity, and rotational state of Asteroid 433 Eros. Icarus 155:3–17
Naeye R (2006) Chandra: taking the universe's x-ray. Sky Telesc 111(1):36–38
Nakamura T, Noguchi T, Tanaka M et al (2011) Itokawa dust particles: a direct link between S-type asteroids and ordinary chondrites. Science 333:1113–1116
Narendranath S, Athiray PS, Sreekumar P et al (2011) Lunar X-ray fluorescence observations by the Chandrayaan-1 X-ray spectrometer (C1XS): results from the nearside southern highlands. Icarus 214:53–66
Narvaez P (2004) The magnetostatic cleanliness program for the cassini spacecraft. Space Sci Rev 114:385–394
Nyren K, Sinnott RW (1992) A tale of two eclipses. Sky Telesc 84(6):678–680
O'Meara SJ, di Cicco D (1993) The night the moon disappeared. Sky Telesc 85(4):107–109, 112
O'Meara SJ (1993) June's colorful lunar eclipse. Sky Telesc 86(5):102
O'Meara SJ (1994) A "Diamond-Ring" lunar eclipse. Sky Telesc 87(3):26–29
O'Meara SJ, di Cicco D (1994) Making sense of November's perplexing lunar eclipse. Sky Telesc 87(6):103–107
Oberg J (2008a) International space station, vol 10, The world book encyclopedia. World Book, Chicago, pp 346a–347
Oberg J (2008b) Mir, vol 13, The world book encyclopedia. World Book, Chicago, pp 613–614
Okada T, Shirai K, Yamamoto Y et al (2006) X-ray fluorescence spectrometry of asteroid Itokawa by Hayabusa. Sky Telesc 312:1338–1341

Ordway FI III, Gardner JP, Sharpe MR Jr (1962) Basic astronautics. Prentice-Hall International, London

Pasachoff JM (2008) Hinode's solar wonderland. Sky Telesc 115(4):64–65

Petersen CC, Brandt JC (1998) Hubble vision, 2nd edn. Cambridge University Press, Cambridge

Peterson V, Jacobs N, M-editors S (1998) Space – new frontiers. Information Plus, Wylie

Pieters CM, Boardman J, Buratti B et al (2009) The moon mineralogy mapper (M^3) on Chandrayaan-1. Curr Sci 96:500–505

Porco CC, Baker BJ et al (2005a) Cassini imaging science: initial results on Saturn's atmosphere. Science 307:1243–1247

Porco CC, West RA, Squyres S et al (2005b) Cassini imaging science: instrument characteristics and anticipated scientific investigations at Saturn. Space Sci Rev 115:363–497

Pyle DM (2000) Sizes of volcanic eruptions. In: Sigurdsson H, Houghton BF, McNutt SR, Rymer H, Stix J (eds) Encyclopedia of volcanoes. Academic Press, San Diego, pp 263–269

Radebaugh J, Lorenz RD, Lunine JI et al (2008) Dunes on Titan observed by Cassini radar. Icarus 194:690–703

Rao J (2008) A multi-hued palette. Sky Telesc 116(1):72–76

Rappaport NJ, Iess L, Tortora P et al (2005) Gravity science in the saturnian system: the masses of Phoebe, Iapetus, Dione and Enceladus. Bull Am Astron Soc 37:704

Redfern G (2009) Lunar fireworks. Sky Telesc 117(6):20–25

Reiter ER (1967) Jet streams. Doubleday, Garden City

Reynolds MD, Sweetsir RA (1995) Observe eclipses, 2nd edn. The Astronomical League, Washington, DC, p 52

Reynolds M, Westfall J (2008) A report on the August 28, 2007 total lunar eclipse. J Assoc Lunar Planet Obs 50(1):26–29

Rhea R (2011) Observing geosynchronous satellites. Sky Telesc 122(4):66–70

Ritchie D, Gates AE (2001) Encyclopedia of earthquakes and volcanoes. Checkmark Books, New York, pp 248–249

Rodell M, Velicogna I, Famiglietti JS (2009) Satellite-based estimates of groundwater depletion in India. Nature 460:999–1002

Roth J (2005) A record-setting solar flare. Sky Telesc 110(3):16–17

Roth J (2006a) NASAs other space telescopes. Sky Telesc 111(1):34–35

Roth J (2006b) GALEX: seeing starbirth, near and far. Sky Telesc 111(1):40–42

Rottman G, Harder J, Fontenla J et al (2005) The spectral irradiance monitor (SIM): early observations. Sol Phys 230(1/2):205–224

Rougier G, Dubois J (1942) Photométrie De L' Éclipse De Lune Du 26 aout 1942. L'Astronomie 56:173–174

Rougier G, Dubois J (1943) Photométrie et colorimétrie de L' éclipse de Lune du 20 Février 1943. L'Astronomie 57:65–66

Rowan L (2005) Cassini drops in. Sky Telesc 307:1222

Saito J, Miyamoto H, Nakamura R et al (2006) Detailed images of asteroid 25143 Itokawa from Hayabusa. Sky Telesc 312:1341–1344

Sánchez-Lavega A (2005) How long is the day on Saturn? Science 307:1223–1224

Schilling G (2009) Shedding new light on dark matter. Sky Telesc 117(4):22–25

Schmude RW Jr (2000) Full-disc wideband photoelectric photometry of the moon. Sky Telesc 95:17–23

Schmude RW Jr (2004) Photoelectric magnitude measurements of the lunar eclipses on May 16, 2003 and Oct. 28, 2004. Ga J Sci 62:188–193

Schmude RW Jr (2008a) Unpublished results

Schmude RW Jr (2008) Uranus, Neptune, and Pluto and how to observe them. Springer Science + Business Media, New York

Schmude RW Jr (2010) Comets and how to observe them. Springer Science + Business Media, New York

Schmude RW Jr, Davies C, Hallsworth W (2000) Wideband photoelectric photometry of the Jan. 20/21, 2000 lunar eclipse. Int Amat Prof Photoelectr Photom Comm 76:75–83

Schneider G, Pasachoff JM, Willson RC (2006) The effect of the transit of Venus on ACRIM's total solar irradiance measurements: implications for transit studies of extrasolar planets. Astrophys J 641:565–571
Schober JH, Schroll A (1973) Photoelectric and visual observation of the total eclipse of the moon of August 6, 1971. Icarus 20:48–51
Seronik G (2000) Full-moon fade-out. Sky Telesc 99(4):116–119
Serway RA (1996) Physics for scientists and engineers, 4th edn. Saunders College, Philadelphia
Shayler DJ (2001) Skylab America's space station. Springer Praxis, Chichester
Shternfeld A (1959) Soviet space science, 2nd edn. Basic Books, New York
Sierks H, Lamy P, Barbieri C et al (2011) Images of asteroid 21 lutetia: a remnant planetesimal from the early solar system. Science 334:487–490
Simkin T, Siebert L (2000) Earth's volcanoes and eruptions: an overview. In: Sigurdsson H, Houghton BF, McNutt SR, Rymer H, Stix J (eds) Encyclopedia of volcanoes. Academic Press, San Diego, pp 249–261
Sinnott RW (1996) Three recent eclipses. Sky Telesc 92(3):101
Sinnott RW (2003) November's two total eclipses. Sky Telesc 106(5):101–105
Sinnott RW (2004) The moon goes dark. Sky Telesc 107(5):80
Sinnott RW, Perryman MAC (1997) Millennium star atlas. Sky Publishing Corporation, Cambridge, MA
(1964) Sky Telesc 27:142–146
(1965) Sky Telesc 29:182–185
(1973) Sky Telesc 46:22–24
(1974) Sky Telesc 48:157
(2005) Sky Telesc 110(3):22
(2006) Sky Telesc 111(1):27
(2008) Sky Telesc 115(3):22
(2010a) Sky Telesc 119(1):14
(2010b) Sky Telesc 119(5):14
(2010c) Sky Telesc 120(3):12
(2010d) Sky Telesc 120(4):16
(2011a) Sky Telesc 121(2):14, 16
(2011b) Sky & Telescope 121(2):18
Souchay J, Kinoshita H, Nakai H et al (2003) A precise modeling of Eros 433 rotation. Icarus 166:285–296
Soulsby BW (2004) Report at http://www.netspeed.com.au/minnah/2004/Summary.html
Soulsby BW (2009) Analysis of lunar eclipse observations. http://www.netspeed.com.au/minnah/LEO97-2x.html
Sparrow G (2009) Spaceflight. DK Publishing, New York
Spudis P, Nozette S, Bussey B et al (2009) Mini-SAR: an imaging radar experiment for the Chandrayaan-1 mission to the moon. Curr Sci 96:533–539
Srama R, Ahrens TJ, Altobelli N et al (2004) The Cassini cosmic dust analyzer. Space Sci Rev 114:465–518
Sreekumar P, Acharya YB, Umapathy CN et al (2009) High energy x-ray spectrometer on Chandrayaan-1. Curr Sci 96:520–525
Stix TH (1981) Plasma waves. In: Lerner RG, Trigg GL (eds) Encyclopedia of physics. Addison-Wesley, Reading, pp 760–763
Stofan ER, Elachi C, Lunine JI et al (2007) The lakes of Titan. Nature 445:61–64
Sutton GP (1992) Rockets and missiles, vol 16, Academic American encyclopedia. Grolier, Danbur, pp 251–260
Tapley BD, Bettadpur S, Ries JC et al (2004) GRACE measurements of mass variability in the Earth system. Science 305:503–505
Tate P (2009) Seeley's principles of anatomy & physiology. McGraw-Hill, Boston, p 468
Taylor GE (1990) Location of geostationary satellites from the British Isles. J Br Astron Assoc 100:90–92

The Hindu: Online Edition of India's National Newspaper (various ISRO press releases), Tamil Nadu, India
Torres G (1986) Space shutle: a quantum leap. Presidio Press, Novato
Tsuchiyama A, Uesugi M, Matsushima T et al (2011) Three-dimensional structure of Hayabusa samples: origin and evolution of Itokawa regolith. Science 333:1125–1128
Tytell D (2005a) Titan: a whole new world. Sky Telesc 109(4):34–38
Tytell D (2005b) Deep impact revisited. Sky Telesc 110(6):16–17
Tytell D (2006) Spitzer: living life to the fullest. Sky Telesc 111(1):44–46
Tytell D (2007) Postcards from Mars and Jupiter. Sky Telesc 113(6):16–17
Vital HC (2003) Brightness of the total lunar eclipse of May 15–16, 2003. Posted at: http://www.geocities.ws/lunissolar2003/Brightness_of_the_Total_Lunar_Eclipse_of_May_15_16_2003.pdf
Vital HC (2011) Observations of the June 15, 2011 total lunar eclipse: a preliminary report. Listed at: http://www.geocities.ws/lunissolar2003/June_2011/Observation_Report.htm
Waite JH Jr, Cravens TE, Ip WH et al (2005) Oxygen ions observed near Saturn's a ring. Science 307:1260–1262
Waite JH Jr, Lewis WS, Kasprzak WT et al (2004) The Cassini ion and neutral mass spectrometer (INMS) investigation. Space Sci Rev 114:113–231
Wasserman LH, Veverka J (1973) On the reduction of occultation light curves. Icarus 20:322–345
Watts RN Jr (1968) NASA's tenth anniversary. Sky Telesc 36:292–293
Weigert A, Zimmermann H (1976) Concise encyclopedia of astronomy. 2nd edn. (trans: Dickson JH). Adam Hilger, London, pp 488–490
Westfall JE (1980) Photographic photometry of the total lunar eclipse of September 6, 1979. J Assoc Lunar Planet Obs 28:116
Westfall JE (1982) Photoelectric photometry of the July 6, 1982 total lunar eclipse. J Assoc Lunar Planet Obs 29:168–171
Westfall JE (1986) The total lunar eclipse of July 6, 1982: a dark and asymmetric umbra. J Assoc Lunar Planet Obs 31:207–212
Westfall JE (1988) Photoelectric photometry of the 1983 JUN 25 partial lunar eclipse. J Assoc Lunar Planet Obs 32:216–221
Westfall JE (1989) Thirty years of lunar eclipse umbrae: 1956–1985. J Assoc Lunar Planet Obs 33:112–117
Whitford-Stark JL (2001) Eruptions. In: Woodhead JA (ed) Earth science, vol II. Salem Press, Pasadena, pp 739–745
Whitiker R (ed) (1996) The nature company guides: weather. US Weldon Owen, Sydney, p 24
Willson RC (1997) Total solar irradiance trend during solar cycles 21 and 22. Science 277:1963–1965
Willson RC, Mordvinov AV (2003) Secular total solar irradiance during solar cycles 21–23. Geophys Res Lett 30(5):1199. doi:10.1029/2002GL016038
Winter FH (1990) Rockets into space. Harvard University Press, Cambridge, MA
Woodbury DO (1958) Around the world in 90 minutes. Harcourt, Brace and Company, New York
Yang Y, Marshak A, Várnai T et al (2010) Uncertainties in ice-sheet altimetry form a space-borne 1064-nm single channel lidar due to undetected thin clouds. IEEE Trans Geosci Remote Sens 48(1):250–259
Young DT, Berthelier JJ, Blanc M et al (2004) Cassini plasma spectrometer investigation. Space Sci Rev 114:1–112
Zhi-Jian Y, Li-Chang L, Yung-Chun L et al (2005) Space operation system for Chang'E program and its capability evaluation. J Earth Syst Sci 114:795–799
Zimmerman R (2000) The chronological encyclopedia of discoveries in space. Oryx Press, Phoenix

Index

R. Schmude, Jr., *Artificial Satellites and How to Observe Them*, Astronomers' Observing Guides, DOI 10.1007/978-1-4614-3915-8, © Springer Science+Business Media New York 2012

T

U

V

W

X

Y

Printed by Publishers' Graphics LLC
MO20120918-2011-185